Energy, Ecology and Environment

Gopal Nath Tiwari

Energy, Ecology and Environment

A Sustainable Nature

Springer

Gopal Nath Tiwari
Indian Institute of Technology Delhi
New Delhi, Delhi, India

BERS Public School
Margupur, Uttar Pradesh, India

ISBN 978-981-99-3996-1 ISBN 978-981-99-3997-8 (eBook)
https://doi.org/10.1007/978-981-99-3997-8

This Springer imprint is published by the registered company Springer Nature Singapore Pte Ltd.
The registered company address is: 152 Beach Road, #21-01/04 Gateway East, Singapore 189721, Singapore

Dedication

This book is dedicated to ***Prof. M.S. Sodha (Guru ji)****, my grandmother (****Late Shrimati Shugia Devi****), and both of my parents (****Late Mrs. Bhagirathi Tiwari*** *and* ***Late Pt. Bashisht Tiwari****), whose philosophies have immensely impacted my life and work and will continue to do so in the future.*

Professor Sodha *was born in Ajmer on February 8, 1932, Rajasthan, pursued his education from UP, including his higher education from Allahabad University, Allahabad (UP), India. This book is dedicated to the celebration of his 92nd birthday in 2024.*

My late grandmother spent her old age in terrible conditions with her mud house getting damaged due to floods. However, one of her practices that I noticed as a child that later helped me with my work was placing a fish in the roots of bottle gourd and pumpkin for better fruiting of the plants. The fish acts as one of best fertilizers, contributing to higher yield, also encouraging good ecology.

My parents' hard work, dedication to provide good education and a healthy environment to their children, belief in God, and conscious decision to shun all unethical practices ensured good morals in my upbringing. They believed that education with hard work is the only way to remove poverty.

Preface

Energy, ecology and environment formed nature globally, and they are strongly connected. Ecology and environment cannot sustain without energy. Solar energy is basic direct source of most of the renewable energy and indirect to non-renewable energy (fossil fuel). Ecology consists of all living organisms including wildlife and humans that needs a good environment consisting of various gases, including carbon dioxide (CO_2), oxygen (O_2), ozone (O_3), carbon monoxide (CO), water vapor (H_2O), sulphur oxides (SO_x), nitrogen oxides (NO_x), Argon (Ar) and Nitrogen (N_2). For a good environment, humans need to use energy at an optimum level.

Nature has provided energy to meet the basic needs of all living organisms for their better survival. Among all living organisms, humans are the most intelligent, and are able to take care of other living organisms. However, humans are responsible for polluting the environment by using non-renewable energy to meet their greed quickly. Non-renewable energy sources (fossil fuel) are very dense in comparison to renewable energy sources, so they release energy quickly, and humans like this. It is also well established that if there is any fast work, there are chances of more damage from that work. Another example, slow cooking keeps all bacteria needed for human for good health with longer life, in comparison to fast or microwave cooking. Hence slow work is always better. For example, if a child starts moving fast, there is a chance of falling to the ground.

In this book, an attempt has been made to explain sustainability of nature by considering various options of using renewable energy for providing better ecological system and environment.

The book has been divided into seven chapters. Nature basically consists of the Sun, Earth, Atmosphere, Environment, etc. and it has been discussed in Chapter 1. In Chapter 2, briefly about solar energy and how solar energy creates various cycles, namely, water/hydrologic (H_2O) cycle, carbon (C), nitrogen (N), oxygen (O), and phosphorous (P) along with wind cycles are discussed. Most of the renewable energy sources suitable for a better and sustainable environment have been discussed in Chapter 3, with their advantages to keep nature sustainable as well. Chapter 4 discusses various methodologies for energy conservation to minimize CO_2 emissions and maximize CO_2 mitigation, as per the Kyoto Protocol and the Paris agreement.

The environment is polluted mainly due to the burning of fossil fuels, and wildfire affects the climate, disturbing the ecosystem, and this is all discussed in Chapter 5. Carbon footprint and net-zero emissions are briefly mentioned in the classification of the ecological system, namely, biotic (living organisms) and abiotic (non-living organisms) in Chapter 6. Further, the important living planet index (LPI) in dealing with population of individual species has been discussed along with the importance of the Sun, atmosphere, and Earth, with respect to ecology. The relevance of renewable energy sources, namely, solar energy, wind, hydropower, biomass, geothermal, OTEC, tidal, wave, etc. with respect to environment along with their advantages and disadvantages are discussed in Chapter 7, the final chapter.

I have taken the help of much literature published in **Wikipedia**, **Byju's**, **Google**, and our book **"Advanced on Renewable Energy Sources"** by G. N. Tiwari and R. K. Mishra with appropriate acknowledgement. If any acknowledgements have been omitted, I express my regret for this, and request all concerned to write in to enable me to rectify the acknowledgements in a revised version in the future.

This book has been primarily written for non-engineering students, faculty, researchers, and entrepreneurs for basic knowledge in energy, ecology, and environment all together in one place. However, engineering students and engineers can also take advantage to teach in B.Tech. and M.Tech. programs.

Examples in each chapter at appropriate section if required are given along with problems and objective questions at end of each chapter. S.I. units and conversion units (Appendix 1) have also been given at appropriate places when it is required. The main physical and chemical constants used are given at beginning of the book.

I am extremely thankful to Ms. Swati Meherishi, Editorial Director, Engineering, Materials Science, Energy, Water, Climate, Springer, New Delhi, India for her support and encouragement during the preparation of manuscript.

I am dedicating this book to my Guru ji, Padama Shri Prof. Mahendra Sodha on his 92nd birthday (February 08, 2024) for his contribution to shape my academic career in Solar Energy since 1977 as a mentor, and I have the privilege to receive unconditional blessing always from him.

I express my gratitude to Late Shri Shyam Prasad Srivastava f/o Dr. Alok Srivastava, Chief Guest, SOALRIS-2024, USA who recently passed away for his continuous blessing for both of us.

In my childhood, my late grandparents and parents had many expectations from me due to many reasons at that time which I never knew. Their blessing was forever with me which I remember today too.

Further, last but not least thanks to Mrs. Kamalawati Tiwari, Mr. Ghanshyam Nath Tiwari, Ms. Ritu Mishra, **Shrivats**, **Shri Ganeshu**, **and Ms. Shrivani (lovely princes)** for keeping patience during these period and helping in so many ways required by me.

New Delhi, India Gopal Nath Tiwari

Contents

About the Author

Prof. Gopal Nath Tiwari (G. N. Tiwari) was born on July 1, 1951, at Adarsh Nager, Sagarpali, Ballia. Dr. Tiwari joined IIT Delhi in 1977 after completing his Ph.D. at Banaras Hindu University as Post-doctoral Fellow (PDF) and became Professor in 1997. To his credit, he has published more than 700 research papers with an h-index of 101. He has published more than 25 books and 20 chapters for various editing books. His work has been recognized by various agencies in terms of Hari Om Ashram Prerit S. S. Bhatnagar Award (ICMRI, Bhavnagar), Vigyan Ratna Award (Government of UP), two times Life Achievement Award (ISAM, USA), European Fellow, Faculty Research Award (Carrier-360) and nominated for International IDEA Award, among other prestigious awards. Dr. Tiwari is Founder Member of SOLARIS based in the UK and has organized three international, and many national, conferences at different venues. He is Chief Editor, Editor, and Reviewer of many international journals with high impact factors. He has supervised more than 100 Ph.D. students in the area of Energy and Environment.

Quotes by Author

Quote-1
"The religion created by human is based on ecological system. It will further start growing again during collapse of Nature by human."

(Gopal Nath Tiwari)

Quote-2
"The blessing is never demanded but pours automatically."

(Gopal Nath Tiwari)

Quote-3
"It took billion years to form the SUN, EARTH, ATMOSPHERE and first living ORGANISM (algae) on the planet Earth which controls the NATURE. Human, one of the most intellectual organisms is destroying the Nature to full fill their greed in couple of century. Save Nature."

(Gopal Nath Tiwari)

Chapter 1
Energy

1.1 Introduction

Energy is broadly classified as renewable and non-renewable. Renewable energy sources are available above the surface of the Earth and are infinite. The main source of all renewable energy is the **Sun**. The non-renewable energy sources are available beneath the ground surface (inside Earth) and are finite; these include coal, petroleum and natural gases. These are also referred to as fossil fuels. The (non-renewable) fossil fuel energy sources are responsible for deteriorating the environment and climate due to emissions of carbon monooxide (CO), carbon dioxide (CO_2) and thermal energy during the burning processes involved in their use. However, renewable energy sources are friendly to the **environment** and mitigate carbon dioxide (CO_2). A clean environment maintains better health of living organisms, including humans and wildlife, comprising the **ecology**.

In next sections, we will discuss briefly about the Sun, Earth and Atmosphere [1, 2].

1.2 Sun

The Sun is a 4.5 billion-year-old star. It is a hot glowing ball of plasma consisting of hydrogen (H_2) and helium (He) at the center of our solar system. The Sun is about 150 million kilometers from the Earth. Living organisms including wildlife and human life (ecology) could not exist on Earth without solar energy. It generates nuclear fusion as its own energy in : its core at temperature of $8 - 40 \times 10^6\text{K}(\approx 1.2 \times 10^7\text{K})$ through

$$H_2 + H_2 \overset{1.2\times10^7\text{K}}{\widehat{=}} He + (\approx 26Mev)44.28 \times 10^{-13}\ \text{J} \tag{1.1}$$

G. N. Tiwari, *Energy, Ecology and Environment*,
https://doi.org/10.1007/978-981-99-3997-8_1

Here, 1Mev $= 1.602 \times 10^{-13}$; Mev stands for mega electron volt

Here, mass of helium (He) [4.00268 Atomic Mass Unit (AMU)] is less than the mass of two hydrogen (H_2) molecules [4.03130 AMU: 1 Atomic Mass Unit (AMU) equals 1.67×10^{-27} kgs]. The difference of mass between two hydrogen (H_2) molecules and one molecule helium (He) is converted into energy. The produced energy is $E = mC^2 = \left((6.7322 - 6.684\,\text{kgs}) \times 10^{-27}\left(\frac{3\times10^8\text{m}}{\text{sec}}\right)^2 = 44.28 \times 10^{-13}\,\text{J}\right)$ as per Einstein theory, at the core of centre of the Sun, which is transferred to the outer surface by convection having black body temperature of 6000 K. Of the energy, 90% is generated between 0 and 0.7R regions at the core ('R' being the radius of Sun). The spectral emission from the outer surface at 6000 K due to Stefan-Boltzmann law has been shown in Fig. 1.1. The emitted radiation is from 0 to infinity consists of ultra-violent $(0 < \lambda < 0.38\,\mu\text{m})$, visible $(0.38\,\mu\text{m} < \lambda < 0.78\,\mu\text{m})$ and infrared radiation $(\lambda > 0.78\,\mu\text{m})$. The energy content in ultraviolent, visible and infrared regions are 6.4%, 48% and 45.6% respectively. Here, it is very important to mention that only energy belonging to the visible range (48%) is useful to living organisms existing on the planet Earth. The emitted radiation from the Sun will be referred to as solar radiation.

Figure 1.2 shows the propagation of solar radiation towards Earth through atmosphere consists of various gases, namely CO_2, O_2, O_3, CO, H_2O, So_x, NO_x and dust. The various physical phenomenons include reflection, absorption and transmission at various levels have also been shown in the same figure. It should be noted that most shortwave length radiation (mostly ultraviolet radiation) which is harmful to all living organism including humans on the planet Earth is reflected back to the extraterrestrial region between the Sun and the top of atmosphere. Further, solar radiation harmful to living organisms, generally infrared and far-infrared, is either absorbed by gases available in the atmosphere or reflected back to the extraterrestrial region. So, only visible radiation reaches the planet Earth, which is essential for survival of living organisms including humans and wildlife. These solar radiations are in the form of either electromagnetic (e/m) waves or photon particles ($E = h\nu$).

The average wavelength of radiation emitted by any hot surface is governed by **Wien's displacement law** as:

$$\lambda_{max}.T = 2897.8 \approx (3000)\,\mu\text{mK} \tag{1.2}$$

The derivation of the above equation is given by Tiwari [3]. Here, T is Kelvin and λ is in μm. The wavelength of radiation is very sensitive to temperature of the surface and inversely proportional to temperature.

Example 1.1 Find out the average wavelength of solar radiation emitted from outer surface of the Sun.

Solution:

As we know, the surface of the Sun = 6000K.

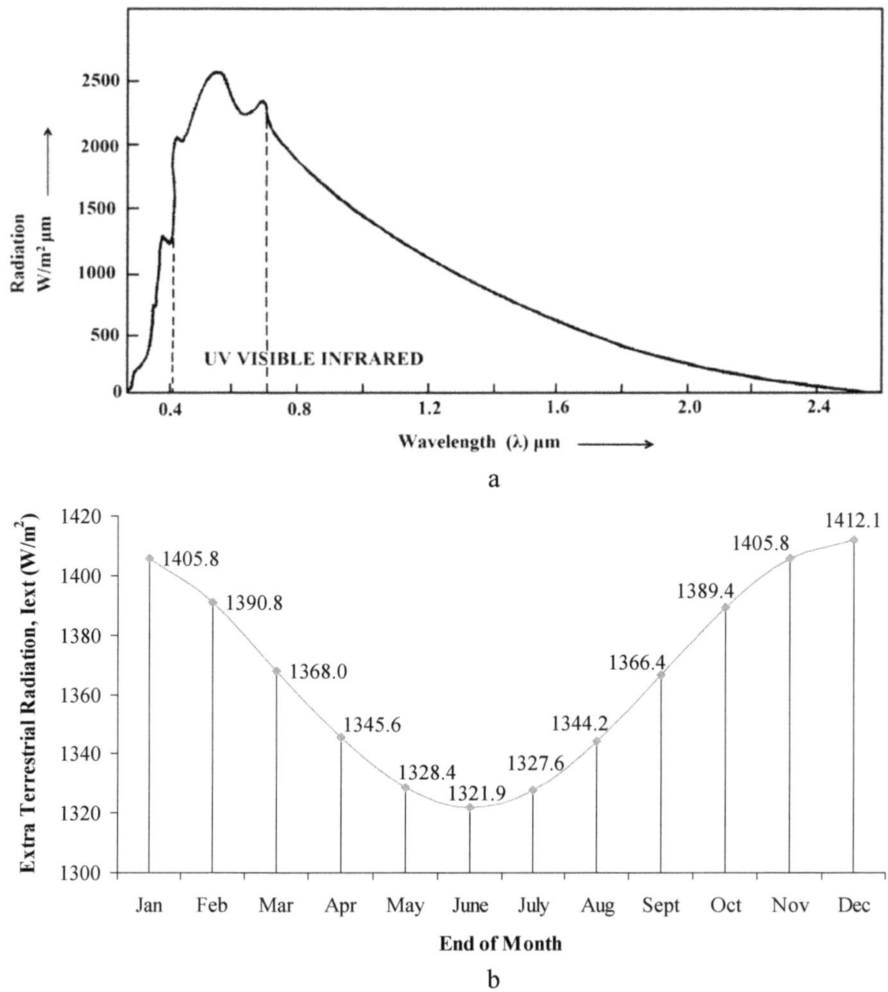

Fig. 1.1 **a** Spectral radiation distribution from outer surface of sun. **b** Variation of I_{ext} with nth day of the year [2]

From Eq. (1.2), one gets, $\lambda = \frac{3000}{6000} = 0.5\mu m$

This wavelength is an average wave length of visible wavelength range shown in Fig. 1.1a.

The **solar constant (I_{sc})** is measured in extra-terrestrial region above top surface of atmosphere and it defined as the solar radiation/flux measured per unit area (W/m^2) normal to surface at mean sun-earth distance. Its numerical value is 1367 W/m^2. The earth moves around the Sun in an elliptical orbit during the year with variation of 1.7 % in distance between sun and earth. It takes 365 days to complete one cycle.

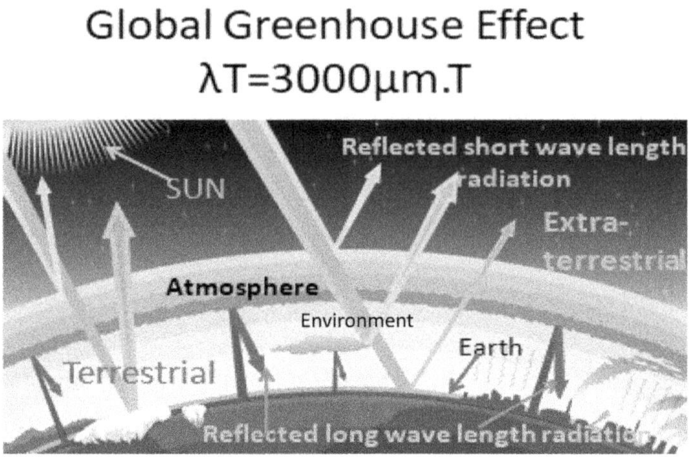

Fig. 1.2 View of solar radiation emitted by the Sun moving towards Earth through atmosphere

Hence, the solar intensity/radiation/irradiance in extra-terrestrial region varies with *n*th day of the year, and its expression is given by [1, 2]

$$I_{ext} = I_{sc}[1.0 + 0.033\cos(360\,\mathrm{n}/365)] \tag{1.3}$$

where the value n should be counted from January 1 of the each year.

The variation of the solar intensity/radiation/irradiance in the extra-terrestrial region with *n*th day of year has been shown in Fig. 1.1b. It can be seen that there is variation of 6.8% in the solar intensity/radiation/irradiance in extra-terrestrial region with respect to its lowest value. Further, the average value is equal to the value of solar constant.

Example 1.2 Find extra-terrestrial solar radiation on June 22 and December 21, 2022 for solar constant of $1367 W/m^2$

Solution:

Given $n = 173$ for June 22, 2020 and $n = 355$ for December 21, 2020.

From Eq. 1.1, one returns:

$$I_{ext} = 1367[1.0 + 0.033\cos(360 \times 173/365)] = 1322.49\,\mathrm{W/m^2}$$

and,

$$I_{ext} = 1367[1.0 + 0.033\cos(360 \times 355/365)] = 1411.43\,\mathrm{W/m^2}$$

One can see that the extraterrestrial solar radiation in June is more than winter due to closeness between Sun and Earth.

1.3 Earth

Earth is the fifth and densest planet in the Solar System with water on its surface unlike the Sun. It is an ellipsoid but almost round with a circumference of about 40,000 km and having a radius (R) of 6,500 km. As mentioned earlier, the Earth rotates in elliptical orbit and takes 365 days to move one round. During its movement in elliptical orbit, the climate on the Earth changes accordingly. Also it spins around its axis and takes 24 hours for one spin, which makes day and night. Since observers are on Earth it seems that the Sun is moving around Earth. Further, the average outer surface temperature of Earth is equal to about 288 K (≈300 K).

Example 1.3 Evaluate the wavelength radiation emitted from ground surface at temperature of $288\,K\,(15°C)$.

Solution:

From Eq. (1.2), one gets, $\lambda = \frac{3000}{300} = 10\,\mu m$

This wave length radiation is very high in comparison with the wave length emitted from out surface of Sun ($10 \gg 0.5$). Hence the solar radiation emitted from outer surface is referred as short wavelength radiation while radiation emitted from the Earth surface is referred as long wave length radiation.

The fossil (remains of blue-green algae and bacteria) has been found in rock/water at least 3×10^9 years ago. The existence of blue–green algae marks the beginning of photosynthesis (living organism) on the planet Earth due to interaction of photon coming from the Sun and H_2O (water). It is the densest planet in the Solar System. The surface area of the Earth is 5.1×10^8 km^2 and has life on its surface. It came into existence some 4.6×10^9 (4.6 billion) years back. About 70% of Earth is covered by water and the remaining 30% is land including forest. Half of the Earth is lit by solar radiation at any one time. It reflects about 1/3rd of radiation back to space between the outer surface of atmosphere and the Sun [3].

The study of the structure of the Earth is known as Geology. It also provides tools to age rocks found, Fig. 1.3. It too describes the histories of rocks found above the Earth [4] and to demonstrate the age of the Earth. Geology has also provided the primary evidence for (i) plate tectonics theory, (ii) the evolutionary history of life, and (iii) the Earth's past climates.

The plate tectonics theory explained as follows:

(a) Earth's outermost layer (lithosphere), Fig. 1.4 which is made up of the crust and upper mantle is broken into large number of rocky plates. These are known as tectonic plates.
(b) These plates move relative to each other from two to fifteen centimeters (one to six inches) per year at different rates,

These plates lie on top of a partially molten layer of rock known as the asthenosphere, Fig. 1.4. The interaction between tectonic plates is responsible for different geological formations such as the Himalaya mountain range in Asia, the East African Rift, and the San Andreas Fault in California (USA), Fig. 1.4.

Fig. 1.3 Earth, mountain and space

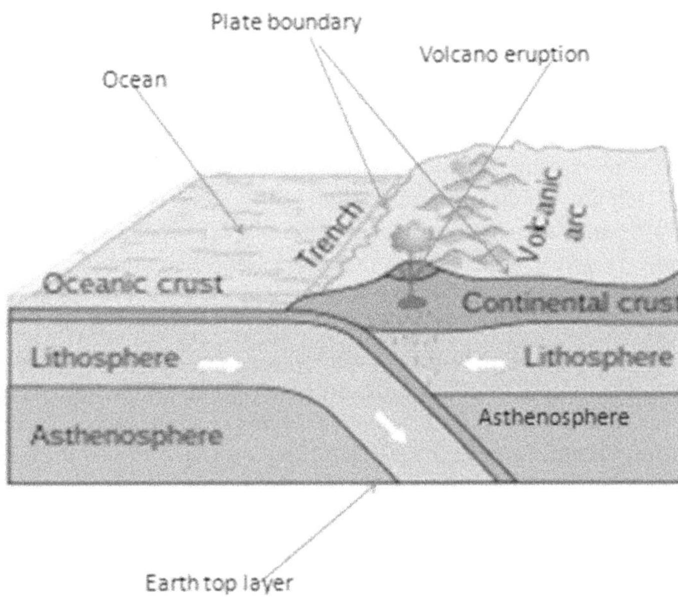

Fig. 1.4 Oceanic-continental convergence resulting in subduction and volcanic arcs

Geologists have been using a wide variety of methods to understand the Earth's structure and evolution which includes field work, rock description, chemical analysis, physical experiments, and numerical modeling etc. It is an important for mineral and hydrocarbon exploration and exploitation, evaluating water resources, understanding of natural hazards, the remediation of environmental problems etc. It also provides insights into past climate change.

Here it is very important to mention that import of minerals/fossil fuels from inside Earth and use of mountains/sand for construction of houses all over the world from one place to another changes the center of gravity of Earth. Due to changes in the center of gravity of Earth, there is disturbance inside the sea which causes the various type of cyclone at different places. This damages the ecology as well as living organisms including humans and wildlife.

Finally Nature is disturbed.

1.4 Atmosphere

The atmosphere is a protective layer of gases that shelters all life on Earth, keeping temperatures within a relatively small range and blocking out harmful rays of sunlight.

The Atmosphere exists between the Sun and Earth. It consists of various gases/ mixture of gases namely CO_2, O_2, O_3, CO, H_2O, SO_x, NO_x, Argon (Ar), Nitrogen (N_2) and dust etc. The particles/molecules presence in atmosphere which scatter solar radiation into terrestrial region are called scatterers and can also include particulates made by human industry. The scattered solar radiation into terrestrial region is known as diffuse radiation and does not have direction like direct radiation. These gases are light weight and erupt from inside the Earth through the plate boundary outwards into space. Being light in weight, they move from the Earth's surface towards the Sun and become stationary due to Earth's gravitational attraction. In other words, an atmosphere is a layer of gas or layers of gases that envelope a planet, and is held in place by its gravity. The atmosphere is a blanket of gases that surrounds Earth consisting of the ocean, land, and ice-covered surface. or other celestial body and composed of about 78% nitrogen (N_2), 21% oxygen (O_2), and one percent other gases [argon (0.9%), carbon dioxide (0.04%) and trace gases (water vapor, ozone, methane, various oxides of nitrogen, neon, and helium)] as shown in Fig. 1.2 [5]. It also minimizes harmful effects of sunlight, ultraviolet radiation (UV), the solar wind, and cosmic rays to protect organisms from genetic damage.

The atmosphere also consists of many layers. It is classified based on altitude and temperature, Table 1.1. These layers are known as **(i) troposphere, (ii) stratosphere, (iii) mesosphere and (iv) thermosphere** [4]. In addition to these, there is the exosphere, which is about 500 km above the Earth's surface. On the planet Earth, the **atmosphere** helps make life possible to all living organism including wildlife and human. The gases and aerosol envelope in atmosphere extends from the ocean, land, and ice-covered surface of a planet earth outward into space.

Following are features of the atmosphere to sustain nature:

(i) It helps to retain the sun's heat of visible wave length range and prevents it from escaping back into extra-terrestrial region
(ii) It protects living organism/life from harmful solar radiation (UV) from the sun.
(iii) It plays a major role in Earth's water cycle.
(iv) It also helps to keep the climate on Earth moderate.

Table 1.1 Characteristic of atmospheric layer from Earth with respect to sea level [4]

Region	Altitude range (km)	Temperature range (°C)	Important characteristics	Remarks
Troposphere	0–11	15 to −56	Weather occurs here	75% of air, 99% of the water vapor and cloud formation
Stratosphere	11–50	−56 to −2	The ozone (O_3) layer is present here	Absorb UV of Sun, warmer than Stratosphere
Mesosphere	50–85	−2 to −92	Meteors burn in this layer	The coldest parts of our atmosphere
Thermosphere	85–800	−92 to 1200	Auroras occur here	Thin layer, Satellites orbit Earth within the thermosphere
Exosphere	500			The air in the exosphere is constant
Ionosphere				Parts of the mesosphere and thermosphere

Without Earth's atmosphere, we can have the following effects:

(i) Any mass such as birds and planes would fall from the sky under gravitational force
(ii) The sky would turn black. Gases and particles in Earth's atmosphere scatter sunlight in all directions as a diffuse radiation. Blue light has more scattering power than other colors because it travels as shorter, smaller waves. This is reason we see a blue sky most of the time.
(iii) There would be no sensation of sound because it requires a medium to travel.
(iv) All the water bodies such as rivers, lakes and oceans would boil away. It would take a billion years to happen. It occurs if the vapor pressure of water in rivers, lakes and oceans exceeds external pressure. In a vacuum, the water readily boils due to direct absorption of solar radiation.
(v) Living organisms would die due to absence of air the atmosphere.

Nowadays, many developed and developing countries are launching satellites from time to time for fast communication from the planet earth to into the exosphere region. During the launching time, it passes through stratosphere (O_3 layer), Fig. 1.5 with discharge of carbon monoxide (CO) and carbon dioxide (CO_2) which depletes the amount of ozone (O_3) in the atmosphere. The carbon mono-oxide (CO) and carbon dioxide (CO_2) converts part of ozone (O_3) into oxygen (O_2) and hence amount of ozone (O_3) is reduced in stratosphere. Due to reduction and depletion of the ozone

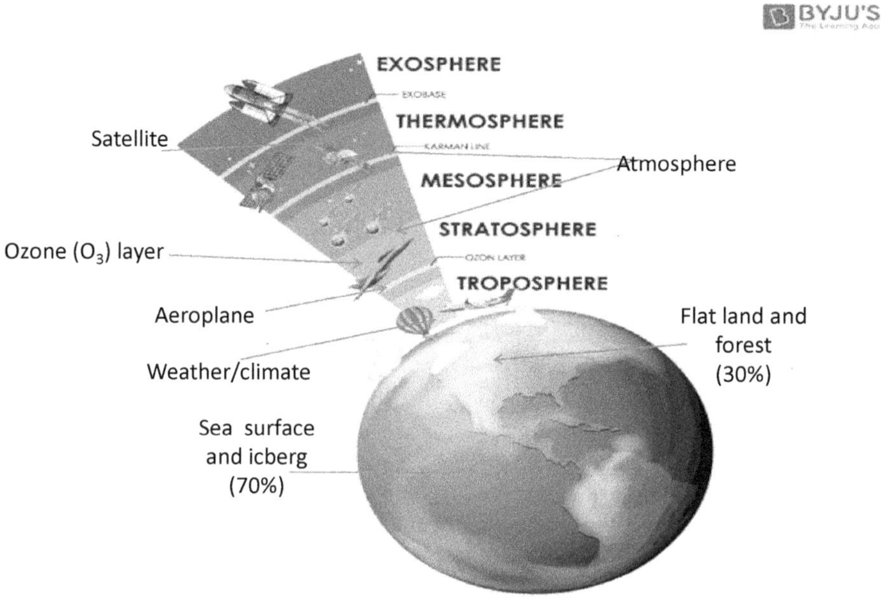

Fig. 1.5 Various layer of atmosphere above the planet Earth [4]

(O_3) ultraviolet (UV) passes from the extraterrestrial region to the terrestrial region, harming living organisms and the ecological system. Finally, it disturbs nature.

Example 1.4 Find out an average temperature of the planet Earth in the absence of atmosphere.

Solution:

In the absence of atmosphere, there is neither the extraterrestrial region nor the terrestrial region. This means a vacuum between the Sun and Earth. The average normal irradiance (beam/direct) will be equal to solar constant ((I_{SC})) i.e. 1367W/m^2 which is directly incident on planet earth and after reflection, it will be absorbed by surface of earth.

The diameter of the Earth (D_e) $= 12.75 \times 10^6$m

The total heat flow in W from the sun to the earth

$$= \text{the projected area of the earth} \times \text{average solar flux (solar constant)}$$

$$\text{or}\, \pi\,(D_e/2)^2 I_{sc} = (\pi/4) \times \left(12.75 \times 10^6\right)^2 \times 1367 = 1.75 \times 10^{14} kW$$

In this case, it is assumed that the rate of incoming solar radiation is completely absorbed by the earth without any reflection and the temperature of earth becomes T_e.

For $\boldsymbol{T_e}$, read the temperature of earth,

The rate outward radiation emitted by Earth

$$= \text{The surface area of the Earth} \times \in \sigma T_e^4 = \pi D_e^2 \in \sigma T_e^4$$

$$= \pi \left(12.75 \times 10^6\right)^2 \times 5.672 \times 10^{-11} \left(kW/m^2K^4\right).T_e^4$$

If incoming average rate solar radiation on earth = the rate outward emitted radiation from the Earth,

Then, $\pi \left(12.75 \times 10^6\right)^2 \times 5.672 \times 10^{-11}.T_e^4 = 1.75 \times 10^{14} kW$

or, $T_e = 278.7K = 5.7°C$, here $\varepsilon = 1$

If the total thermal energy coming from the sun to the earth is multiplied by 0.7 due to the Earth's albedo (reflection from earth), then.

Earth's temperature, $T_e = 195K = -78\,°C$

This shows that existing atmosphere between the sun and the earth blocks maximum outgoing long wave length radiation so that the average temperature of earth has been raised to about 15°C(288K) Example 1.8. In this case, it will be a frozen world at beginning. It might have been happened 2.4 to 2.3 billion years ago with the entire planet frozen, known as "Snowball Earth". As time passes without atmosphere, the earth temperature will rise without atmosphere and after some time frozen water will start melting and then become hotter and may boil.

1.4.1 Climate

The climate of any region/location of country is determined on the basis of long term (≈ average over 30 years time) weather condition parameters namely (i) Mean Monthly Temperature (MMT), (ii) Mean Monthly Relative Humidity (MMRH) (%), (iii) Mean Monthly Precipitation, MMP (mm) (rain/snow fall), (iv) Number of average clear days (Blue sky condition), (v) Mean Atmospheric Pressure (MAP) and (vi) Wind etc. The most widely used classification scheme over the world was the **Köppen climate classification** [6].

The Köppen climate has been classified into five main climate groups. Each group is further divided based on seasonal precipitation/rainfall/snowfall and temperature patterns. The five main groups are named as:

(i) A (tropical humid climate): The average temperature of the coldest month ≥ 18° C. It has no dry season.
(ii) B (arid/dry/semi-arid): Potential evaporation exceeds precipitation due to low relative humidity
(iii) C (warm temperature climate): The average temperature of the coldest month of the (Mid-latitude) climates years falls in between 18°C > –3 °C (more than –3 °C but less than 18°C)
(iv) D (continental/cold snow forest): The average temperature of the coldest month is ≤–3° C and
(v) E (polar/cols climate): Average temperature for all months < 10° C

Table 1.2 Criteria for the classification of climates [5]

Climate	Mean monthly temperature	Relative humidity (%)	Precipitation (mm) (rain/ snow fall)	Number of clear days	Examples
Hot and dry (HD)	>30	<55	<5	>20	Jodhpur, Saudi Arabia
Warm and humid (WH)	>30	>55	>5	<20	Bombay, Pacific countries
Moderate (MO)	25–30	<75	<5	<20	Bangalore
Cold and cloudy (CC)	<25	>55	>5	<20	Srinagar, Europe
Cold and sunny (CS)	<25	<55	<5	>20	Leh, Singapore
Composite climate (CO)	This applies when six months or more do not fall within any of the above categories.				New Delhi

In addition to five, there is additional classification as H known as Highlands (hilly area with snow covers) which has cold condition due to elevation. Each group and subgroup is represented by a letter.

It is to be noted that there is basic three main parameters namely temperature, relative humidity and precipitation to classify climate. This is mainly happening in countries like European countries but this basis cannot be applied in another area which has variation in (a) number of average clear days (solar radiation/blue sky condition), (b) mean atmospheric pressure (MAP) and (c) wind etc.

For example, the Indian climate has more additional parameters in addition to Köppen climate classification like number of clear days/sunny condition/blue sky condition. After considering these parameters, the Indian climate has been classified into six categories as given in Table 1.2. However, there is similarity between two classifications with differences only of temperatures range. For example, Hot and dry (HD), warm and humid (WH), cold and cloudy (CC) and cold and sunny (CS) are similar to C (warm temperature climate), A (tropical humid climate), D (continental/ cold snow forest).

1.4.2 Weather

Weather is the daily atmospheric conditions for a given specific place/location to know its effects on the everyday life and activities of humans [6]. The atmospheric conditions, namely the brightness/blue sky condition and sunshine hours, wind, visibility, humidity, cloudiness, air temperature, precipitation/rain/snowfall etc. measure the weather of a particular day and time of location/place It can be changed within

Table 1.3 The difference between climate and weather [6]

S.no	Climate	Weather
1	Climate is statistical average weather condition over a period of over 30 years of a given location/place	Atmospheric changes based on day to day information at specific time is weather
2	Climate changes over a long period	Weather condition changes frequently
3	Climate significantly impact industries, agriculture, livelihood of inhabitants	Weather can impact day to day human life like transportation, agriculture, communication and agriculture
4	It mainly depends on temperature, wind, relative humidity etc	In this case, in addition to temperature, wind and relative humidity, air pressure, rain, cloudiness, snow, storms, precipitation is also conidered

minutes, or hours in a given climatic conditions. These changes mostly happen in the nearest atmospheric layer namely troposphere to the ground as shown in Fig. 1.5. Nowadays, weather forecast has become very easy due to satellite observation.

Due to sudden changes in weather conditions, the following effects have been observed:

(i) Many crops/plants are damaged due to heavy rains along with strong winds/storms leads to food shortage.
(ii) Loss of human/animal life due to strong lighting in the atmosphere.

The some of difference between climate and weather has been given in Table 1.3.

1.4.3 Season

The season is determined by the changes in weather during a specific time of the year. Contrastingly, the climate is the behavior of the atmosphere over a longer timeline. Thus, the prime difference between season and climate is also the measure of duration.

1.5 Global Greenhouse Effect

Atmosphere as mentioned in Sect. 1.4 has very unique two properties:

(i) It transmits only **short wave length radiation** ($0.5\mu m$), Example 1.1 coming from Sun between 0.26 μm ($\approx$0.3 μm) to 2.6 μm ($\approx$3 μm) and
(ii) It behaves as a opaque for the rest of radiation having wave length either below 0.26 μm ($\approx$0.3 μm) or above 2.6 μm ($\approx 3\mu$m). In another words it behaves as a opaque for long wave length solar radiation (Example 1.3)

Sometimes, the atmosphere is also referred to as a filter of required solar radiation coming from Sun for living organisms. The atmosphere can also be referred to as a **transparent layer** to short wavelength of solar radiation and opaque to long wavelength radiation. The regions between the Sun and the Atmosphere and between the Atmosphere and the planet Earth are referred to as extraterrestrial and terrestrial regions respectively.

There is strong relation between the Sun, the Atmosphere and the planet Earth for survival of all living organism on the planet Earth. For example, the short wave length radiation coming from the Sun is partially reflected, absorbed and transmitted through the atmosphere and reaches the Earth as shown in Fig. 1.2. The Earth absorbed a part of short wave length radiation after reflection from the Earth, and then Earth's temperature increases and it emits long wavelength radiation due to Wien's displacement law ($\lambda T = 3000\,\mu\text{mK}$). This long wavelength radiation (thermal energy) emitted from the Earth is blocked by the atmosphere due to its second property, as mentioned above. Hence, the air in the terrestrial region is heated fast due to its low heat capacity and the temperature of air rises. Now, if the temperature above ice of oceanic continent rises, the ice starts melting and forms water which is in a steady state. We also consider the radiation coming from Sun in the form of photon ($E = h\nu$) contains energy much higher than the binding energy of H_2O. If the photon strikes H_2O then Hydrogen (H_2) and Oxygen (O_2) is separated. After that Oxygen (O_2) is released to environment to sustain a clean environment (209.46 ppm) which is the basic need of human being for survival. Further, the Hydrogen (H_2) reacts with carbon dioxide (CO_2) of the environment (412 ppm) to make it hydrocarbon (mass) in the form of **algae (biomass)**. So, the algae were the first living organisms on the planet Earth. This reduces the level of carbon-dioxide (CO_2) in the environment. This process provides additional O_2 and reduces CO_2 in environmental to sustain it. The green plant which contains maximum H_2O consumes CO_2 released by human beings, industries and the environment, and the plants release O_2 for living organisms including human beings for good health as shown in Fig. 1.6. This also helps to clean environment and process, and is known as photosynthesis. Since the color of algae is green and atmosphere and earth under consideration acts as roof and floor hence this phenomenon is known as global Green House Effect. From algae formation to human being, it might have taken many million years. It is a history [7].

The chemical reaction of photosynthesis between plant and photon of solar radiation is given by.

$$6CO_2 + 6H_2O + \text{solar radiation (Photon, h }\nu) \rightarrow C_6H_{12}O_6 + 6O_2 \tag{1.4}$$

1.6 Global Warming

As we are aware that the natural level of carbon dioxide (CO_2) in the environment is estimated to be about 280 ppm per volume pre-industrial level due to the respiration of plants and animals. This level of CO_2 existed steadily for centuries before the

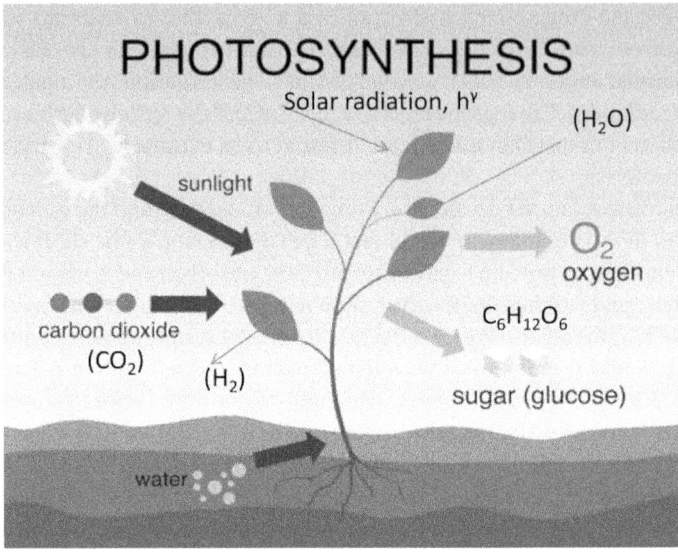

Fig. 1.6 Photosynthesis process between plant and photon of solar radiation

industrial revolution (WW-II). Nowadays, the major concern is the possible disturbing of the Earth's energy balance leading to an increase in the average global air temperature.

The following equation provides the simplest form of the Earth's energy balance [8] and is written in a steady condition:

$$I_{sc}(1 - f) \times \pi R^2 = \epsilon \sigma T^4 \times 4\pi R^2 \tag{1.5a}$$

$$\left[\begin{array}{c}\text{The rate of solar radiation reaching outer surface} \\ \text{of Earth equivalent to its cross section at center}\end{array}\right] = [\text{The rate of solar radiation emitted from outer surface of Earth}]$$

or

$$T = \left[\frac{I_{sc}(1 - f)}{4\epsilon\sigma}\right]^{1/4} \tag{1.5b}$$

where I_{sc} is the solar constant (1367 W/m^2), f is the albedo or reflectivity of the Earth (0.31), R is the Earth's radius (6500 km). The ϵ is the effective emissivity of the Earth for infrared radiation to outer space (0.615), σ is the Stefan-Boltzmann constant ($5.5597 \times 10^{-8}\ Wm^{-2}K^4$, T is the absolute global average air temperature.

Here, it important to note that atmospheric gases affect both albedo/reflectivity (f) and the effective emissivity (ϵ).

With increases in aerosols and cloudiness, increases CO_2 level and decrease in effective emissivity of the Earth (ϵ). We probably error on the pessimistic side (i.e., predicting a temperature change that is too large) if we assume, as we shall here, that an increasing CO_2 level works only to decrease effective emissivity of the Earth (ϵ).

The following expression gives a reasonable estimate for variation of effective emissivity of the Earth (ϵ) with respect to CO_2 concentration as

$$\epsilon = 0.642 - (8.45 \times 10 - 5)\text{pCO}_2 \tag{1.6}$$

where pCO_2 is the concentration of carbon dioxide (CO_2) in the atmosphere in parts per million by volume (ppmv).

Example 1.5: Find out the energy associated with one photon which has a wavelength λ of 700 nm (red end of visible solar spectrum). Known 1 nm = 0.001 $\mu m = 10^{-9} m$.

Solution: The energy associated with the photon (E) is given by

$$E = h\nu$$

$$= \frac{hc}{\lambda}, \text{ where c is the velocity of light.}$$

For h = 6.63×10^{-34} J.sec and c = 3×10^8 m/sec, we have

$$E = \frac{6.63 \times 10^{-34} J \cdot sec \times 3 \times 10^8 m/sec}{700 \times 10^{-9} m}$$

$$= 2.84 \times 10^{-19} J$$

The value of E in terms of eV is given by

$$E = \frac{2.84 \times 10^{-19}\,\text{J}}{1.60 \times 10^{-19}\,\text{J/eV}} = 1.7\text{eV};\ 1\text{eV} = 1.6021 \times 10^{-19}$$

as 1 eV = 1.6×10^{-19} J.

Example 1.6 Calculate the number of fusion reaction equivalent to total energy generated from outer surface of Sun.

Solution:

$$\text{The rate of outward solar radiation emitted by Sun's surface}$$
$$= \text{the surface area of the Sun} \times \varepsilon\sigma T_s^4, \text{W}$$

where $\varepsilon = 0.9$, $\sigma = 5.672 \times 10^{-11} \mathrm{W/m^2K^4}$ and T_s can be considered as 6000 K.

The total energy emitted by the Sun in 24 h, E_{ST} = The surface area of the Sun $\times$ $\varepsilon\sigma T_s^4 \times 24$W h.

Here, radius of the Sun $= 6.957 \times 10^8$ m, then

In one fusion reaction energy generation, $E_f = 44.28 \times 10^{-13}\mathrm{J} = 44.28 \times 10^{-13} \times 0.00028 = 1.22 \times 10^{-15}$ Wh (Eq. 1.1).

So, the number of fusion reaction took place $= \frac{1.69925 \times 10^{24}}{1.22 \times 10^{-15}} = 1.3928 \times 10^{39}$

Example 1.7 Calculate effective emissivity of the Earth(ϵ) with respect to CO_2 concentration of 280 ppm.

Solution: From Eq. 1.6, an expression for effective emissivity of the Earth(ϵ) with respect to CO_2 concentration is given by

$$\epsilon = 0.642 - (8.45 \times 10 - 5)\mathrm{pCO_2}$$

Here, CO_2 concentration = 280 ppm (given). Substitute this value in above equation then

$$\epsilon = 0.642 - (8.45 \times 10 - 5) \times 280 = 0.642 - 0.02366 = 0.618$$

Example 1.8 Calculate global average atmospheric air temperature prior to industrialization.

Solution:

In order to get global atmospheric air temperature, substitute an appropriate numerical value in Eq. 1.5b then

$$T = \left[\frac{I_{sc}(1-f)}{4\epsilon\sigma}\right]^{\frac{1}{4}} = \left[\frac{1367(1-0.31)}{4 \times 0.615 \times 5.5597 \times 10^{-8}}\right]^{\frac{1}{4}}$$
$$= \left[\frac{1367(1-0.31)}{4 \times 0.615 \times 5.5597 \times 10^{-8}}\right]^{\frac{1}{4}}$$
$$= [68.96 \times 10^8]^{\frac{1}{4}} = 288K = 15^\circ\mathrm{C}$$

Problems

Problem 1.1 Evaluate an average wavelength radiation emitted from surface at temperature of 100 °C and 200 °C.

Hint: See Example 1.1

Problem 1.2 Evaluate an extra-terrestrial solar radiation on February, 29 and June 22, 2022 (leap year) for solar constant of $1367\text{W}/\text{m}^2$

Hint: See Example 1.2

Problem 1.3 Calculate total radiation from the Sun for given radius (R) of $12.75 \times 10^6\text{m}$ and temperature of 6000 K.

Hint: See Example 1.4

Problem 1.4 Find out the energy associated with one photon having frequency of 0.3 and 3 μm

Hint: See Example 1.5

Problem 1.5 Calculate $E = mc^2$ for fusion taking place inside Sun.

Hint: See Eq. (1.1)

Problem 1.6 Plot the curve between energy (E) and λ which is varying between 0.3 and 3 μm (visible wavelength).

Hint: See Example 1.5

Problem 1.7 Plot the curve between frequency (ν) and λ which is varying between 0.3 and 3 μm

Hint: See Example 1.5

Problem 1.8 Plot the curve between wavelength (λ) in hertz and frequency (ν) in μm which is varying between 0.3 and 3 $\mu\text{m}\left[1\mu\text{m} = 2.997 \times 10^{14}\text{Hz}\right]$ and λ which is varying between 3 and 0.3

Hint: See Example 1.5

Problem 1.9 Plot the variation of effective emissivity of the Earth(ϵ) with respect to CO_2 concentration vary from 280 to 420 ppm.

Hint: See Example 1.7

Problem 1.10 Plot global average atmospheric air temperature with varying effective emissivity of the Earth(ϵ).

Hint: See Example 1.7 and 1.8.

Objective Questions

1.1 The wavelength radiation emitted from surface is

(a) Proportional to temperature
(b) Inversely proportional to temperature
(c) Equal to temperature
(d) None of them

Answer: (b).

1.2 Solar constant is measured in/on

(a) Surface of earth
(b) Terrestrial region
(c) Extraterrestrial region
(d) The surface of Sun

Answer: (c).

1.3 The maximum radiation is emitted by

(a) The surface of sea
(b) The surface of Earth
(c) The surface of rock
(d) The surface of Sun

Answer: (d).

1.4 The sun emits

(a) Short wavelength radiation
(b) Long wavelength radiation
(c) Zero wavelength radiation
(d) None of them

Answer: (a).

1.5 The Earth emits

(a) Short wavelength radiation
(b) Long wavelength radiation
(c) Zero wavelength radiation
(d) None of them

Answer: (b).

1.6 Atmospheric gases with respect to air are

(a) Zero weight
(b) Heavy weight
(c) Light weight
(d) None of above

Answer: (c).

1.7 Ozone (O_3) layer is disrupted by

(a) Satellite
(b) Aero plane
(c) Dust
(d) All of them

Answer: (a).

1.8 Frequency (ν) is

(a) Inversely proportional to wavelength (λ)
(b) Proportional to wavelength (λ)
(c) Equal to wavelength (λ)
(d) None of them

Answer: (a)

1.9 Wavelengths (λ) is

(a) Inversely proportional to frequency (ν)
(b) Proportional to frequency (ν)
(c) Equal to frequency .
(d) None of them

Answer: (a)

1.10 Energy emitted by surface of Sun at 6000 K is

(a) Higher than energy generated at core of Sun
(b) Lower than energy generated at core of Sun
(c) Equal to energy generated at core of Sun
(d) None of them

Answer: (b).

1.11 The main greenhouse gases of atmosphere (stratosphere) are

(a) Nitrogen (N_2), oxygen (O_2)
(b) Argon, carbon dioxide (CO_2)
(c) Trace gases namely vapor, ozone(O_3),, methane, various oxides of nitrogen, neon, and Helium (He)
(d) All of them

Answer: (d).

1.12 The algae is formed with interaction of photon

(a) Water molecule of free moving water surface
(b) Water molecule of free stagnant water surface
(c) Water content in the plant
(d) All of them

Answer: (b) and (c).

1.13 The effective emissivity of the Earth (ϵ) will increase

(a) With increase of CO_2 concentration
(b) With decrease of CO_2 concentration
(c) With no effect of CO_2 concentration
(d) None of them

Answer: (b).

1.14 The effective emissivity of the Earth (ϵ) will decrease

(a) With increase of CO_2 concentration
(b) With decrease of CO_2 concentration
(c) With no effect of CO_2 concentration
(d) None of them

Answer: (a).

1.15 The global average atmospheric air temperature will increase

(a) With increase of CO_2 concentration
(b) With decrease of CO_2 concentration
(c) With no effect of CO_2 concentration
(d) None of them

Answer: (a).

References

1. J.A. Duffie, W.A. Beckmann, *Solar engineering of thermal processes* (John Wiley and sons, New York, 1991)
2. G.N. Tiwari, Arvind and Shyam, *Handbook of Solar Energy*, Sringer (2016)
3. G.N. Tiwari, *Solar Energy: Fundamental, Design, Modeling and Application* (Narosa Publishing House PVT. Ltd., New Delhi, 2002)
4. Definition, Layers of Atmosphere, Composition of ...—Byju's (2020). https://byjus.com › physics › atmosphere
5. Difference between Weather and Climate – Vedantu https://www.vedantu.com › biology › difference-betwe...
6. Tiwari, G.N., Ecology, Environment and Human Being Ecol Conserv Sci 1(3): ECOA.MS.ID.555565 (2020). https://doi.org/10.19080/ECOA.2020.01.555565
7. Carbon cycle modelling: illustrations of ...https://maths.anu.edu.au›cmaprocvol25-enting
8. "Earth's Atmosphere Composition: Nitrogen, Oxygen, Argon and CO_2". *Earth How*. 2017–07–31. Retrieved 2019–10–22.

Chapter 2
Solar Energy and Various Cycle

2.1 Introduction

As mentioned in previously in Chap. 1, solar energy is emitted from the Sun. It is always available in the terrestrial zone, and numerical values vary from 1405.8 to 1412.1 W/m^2 with minimum value of 1321.9 W/m^2. Its average numerical value is 1367 W/m^2 $\left(\frac{1412.1+1321.9}{2} \text{W/m}^2\right)$ in extra-terrestrial region of Fig. 1.1b which is equal to solar constant. The radiant energy from the Sun is intermittent in terrestrial region due to some absorption, reflection and scattering by atmosphere. The maximum value of solar energy available on the planet Earth will be less than the numerical value of the solar constant (1367 W/m^2) due to reflection from the outer surface of atmosphere, atmospheric particles and Earth's surface (Fig. 1.2), and absorption by the atmosphere. The maximum value at high altitude (hilly region) on the planet Earth has been reported as about 1100 W/m^2 and on flat land with clear sky conditions as 1000 W/m^2 during sunshine hours at noon, and zero during dark. So an average value of solar radiation per day may be considered as 500 W/m^2. Currently, living organisms, especially humans, need 24-hour energy to meet requirements. For human requirements, solar energy needs to be stored for use during the night.

Further, most of solar equipment including solar cells and PV modules should be tested in outdoor conditions. However, climatic condition vary from country to country due to different latitude and altitudes and it becomes difficult to compare the performance of solar equipment developed by different countries. So it was decided to test all solar equipment under international standard test condition (IST) for indoor condition by simulation. For this, solar intensity and surrounding temperature has been fixed as 1000 W/m^2 and 25 °C. The solar intensity of 1000 W/m^2 has been fixed due to its maximum value available on flat land, and surrounding air temperature of 25 °C has been considered as equivalent to an average value of global ambient air temperature in troposphere. It may be due to fact that the maximum value of ambient

G. N. Tiwari, *Energy, Ecology and Environment*,
https://doi.org/10.1007/978-981-99-3997-8_2

55 °C, and minimum temperature at ground level reaches around 0 to −5 °C, hence average value is considered as 25 °C.

2.2 The Hydrologic Cycle

The hydrologic cycle is also referred as water (H_2O) cycle as well as the hydrological cycle [1]. It describes the continuous movement of water on above (Earth-atmosphere system) and below the surface of the Earth. In this case, water changes its state from one phase to another for example from ice (solid) to water (liquid) or water (liquid) to vapor/steam (gaseous). Before the start of life on the planet earth, about 70% of land was covered by snow (see Example 1.4). However, the total number of water particles (molecules, H_2O) remains the same during change of state. In other words, if it is possible to collect and boil 1 kg of water, it will still retain a mass of 1 kg as vapor/steam and vice-versa form ice to liquid.

There are main four stages of hydrologic cycle. These are as follows.

(a) **Evaporation**

As mentioned in Chap. 1, there existed a snowball Earth which was partially melted due to the greenhouse effect, and the first living organism (algae) was created. It might have taken a billion years to have much water on the planet Earth. It is well known fact that this was happened due to thermal/heat energy (e/m waves) emitted from the Sun because the Sun is the ultimate source of all renewable energy directly and non-renewable energy indirectly. The available thermal/heat energy on water surface raises its temperature after absorption. When the water pressure becomes higher than surrounding air pressure, there is diffusion of water molecules (H_2O) into atmosphere which increases the relative humidity (water content in atmosphere). Higher the relative humidity (large percentage of water in the atmosphere), lower is water molecule transfer into atmosphere. This is referred as evaporation. Water vapor pressure depends on temperature (Table 2.1). In other words, evaporation generally happens when water molecules at the surface of water bodies become excited and rises into the air. These molecules with the highest kinetic energy accumulate into water vapor clouds. Evaporation usually takes place below the boiling point of water. The same process is called **evapotranspiration,** if evaporation takes place on the leaves of plants it is also known as **respiration** from the plant.

Example 2.1 Evaluate the rate of evaporative in $\mathrm{W/m^2}$ and heat transfer coefficient in $\mathrm{W/m^2}$ °C from water surface (35 °C) to an ambient air temperature (15 °C) with a relative humidity of 50%.

Solution

The vapor pressure $P(T)$ in $\mathrm{N/m^2}$ at any temperature T (in °C) can be calculated by using either following expression [2] or steam Table 2.1.

Table 2.1 Steam table for saturation vapor pressure

Temp. (K)	P (N/m²)	Temp. (K)	P (N/m²)	Temp. (K)	P (N/m²)
273	610.8	304	4491.0	334	20860.0
274	656.6	305	4743.0	335	21840.0
275	705.5	306	5029.0	336	22860.0
276	757.6	307	5318.0	337	23710.0
277	812.0	308	5622.0	338	25010.0
278	871.8	309	5940.0	339	26150.0
279	934.5	310	6274.0	340	27330.0
280	1001.2	311	6624.0	341	28560.0
281	1072.0	312	6991.0	342	29840.0
282	1147.2	313	7375.0	343	31160.0
283	1227.0	314	7777.0	344	32530.0
284	1311.6	315	8198.0	345	33960.0
285	1401.4	316	8639.0	346	35430.0
286	1496.5	317	9100.0	347	36960.0
287	1597.3	318	9583.0	348	38550.0
288	1703.9	319	10,086.0	349	40190.0
289	1816.8	320	10,612.0	350	41890.0
290	1936.2	321	11,162.0	351	43650.0
291	2062.0	322	11,736.0	352	45470.0
292	2190.0	323	12,335.0	353	47360.0
293	2337.0	324	12,961.0	354	49310.0
294	2485.0	325	13,613.0	355	51350.0
295	2642.0	326	14,340.0	356	53420.0
296	2808.0	327	15,002.0	357	55570.0
297	2982.0	328	15,641.0	358	57800.0
298	3166.0	329	16,511.0	359	60110.0
299	3360.0	330	17,313.0	360	62490.0
300	3564.0	331	18,147.0	361	64950.0
301	3778.0	332	19,016.0	362	67490.0
302	4004.0	333	19,920.0	363	70110.0
303	4241.0				

$$P(T) = \exp\left[25.317 - \frac{5114}{T + 273}\right] \tag{2.1}$$

Thus, the vapor pressures at wetted and ambient air temperatures can be calculated as,

$$P(T_w) = \exp\left(25.317 - \frac{5144}{273 + 35}\right) = 5517.6\,\text{N/m}^2$$

and

$$P(T_a) = \exp\left(25.317 - \frac{5144}{273 + 15}\right) = 1730\,\text{N/m}^2$$

Using $h_c = 2.8 + 3$ V, W/m^2 °C [3] and If V = 0, one can have the rate of evaporation in W/m^2 as:

$$\dot{q}_{\text{ew}} = 16.273 \times 10^{-3} \times 2.8 \times (5517.6 - 0.5 \times 1730) = 211.99\,\text{W/m}^2$$

The evaporative heat transfer coefficient, which is the rate of evaporation per unit temperature difference, can be calculated as

$$h_{\text{ew}} = \frac{\dot{q}_{\text{ew}}}{T_w - T_a} = \frac{211.99}{35 - 15} = 10.60\ \text{W/m}^2\,°\text{C}$$

(b) **Sublimation**

If the snow or ice changes directly into water vapor without changing its state into water, it is referred as sublimation. It is generally happens if dry winds are flowing above it with low relative humidity. Sublimation can be observed on mountain and hill peaks, where the air pressure is quite low due to low air temperature (Table 2.1). The low air pressure helps to sublimate the snow into water vapor as less energy is utilized in this process. On the planet Earth, the basic source of sublimation is from the snow/ice sheets covering its poles.

(c) **Cloud Formation**

The water vapor evaporated from water surface moves in upward direction, Example 2.1 and accumulated in the troposphere of atmosphere which is eventually cools down due to the low temperatures found at high altitudes. These vapors become tiny droplets of water and ice which finally form clouds.

(d) **Vapor Precipitation (VP)**

The cloud which consists of vapors formed from water surface of ground will only condense into water droplets, if its temperature is above zero degrees centigrade (>0 °C). However, it cannot condense without contact of dust/other impurities. Hence, water vapors attach itself with particle's surface. When large number of droplets merge, it falls out of the clouds under gravity on the ground surface. This process is referred as precipitation/rainfall. In particularly cold weather or extremely low air pressure, the water droplets freeze and fall as snow. In addition to above process, there are infiltration and runoff of falling water/snow/ice from the cloud. The precipitation is measured in millimeters (mm). The rain fall over the course of 1 h, 1 day, 1 month or 1 year is measured for a given area of water container. It is very easy to measure rain fall. The precipitated/rain water can be used for the following.

(i) **Potable water**: It is used for drinking purpose. The rain water should be collected for drinking after 2/3 rain periods so that harmful particles/particulates presence in the atmosphere due to air pollution can be filtered from the environment. It can also be used for other domestic applications.
(ii) **Agriculture**: In this case, suitable water ponds should be prepared to store rain water nearby agricultural land so that collected rain water can be used for irrigation in the emergency and drought situations. The same water pond can also be used for aquaculture to make the water pond more economical. This has been practiced in Thailand. It is friendly to the environment and climate without using any fossil fuels.

The equation used to evaluate precipitation is simple and can be written as

$$\text{Volume of Precipitation (VP)} = \text{Roof/pond area (Length} \times \text{Width)} \times \text{Rainfall} \tag{2.2}$$

Example 2.2 Calculate amount of water collected through roof of area 3 m × 3 m with rail fall/precipitation of 20 mm.

Solution

From Eq. 2.2, one has the following.

$(\text{VP}) = \text{Length} \times \text{Width} \times \text{Rainfall} = 3\text{ m} \times 3\text{ m} \times 20\text{ mm} = 3\text{ m} \times 3\text{ m} \times 0.02\text{ m} = 0.18\text{ m}^3$

(e) **Infiltration (G)**

Rainwater gets absorbed into the ground through the process of infiltration due to porosity of soil of ground. The level of absorption varies based on the soil material and the water has seeped inside ground. For instance, rocks will retain comparatively less water than soil due to high density. Ground surface water can either follows streams or rivers or pond on hills. But sometimes, water falling on the ground might just sink deeper and form aquifers due to stored thermal energy inside the ground.

(f) **Runoff (R)**

If the water from rainfall lands on mountains or hills, then it does not form aquifers due to high density of the mountains or hills, but it roll down under gravity and finally flowing down the sides of mountains and hills and either becomes water pond or forming rivers. This process is called runoff/flowing water stream. In colder regions, icecaps form when the amount of snowfall is faster than the rate of evaporation or sublimation.

The whole hydrologic cycle has been shown in Fig. 2.1a. The physical properties of runoff water are given in Table 2.2.

Basically, water balance equation, Fig. 2.1b is given by

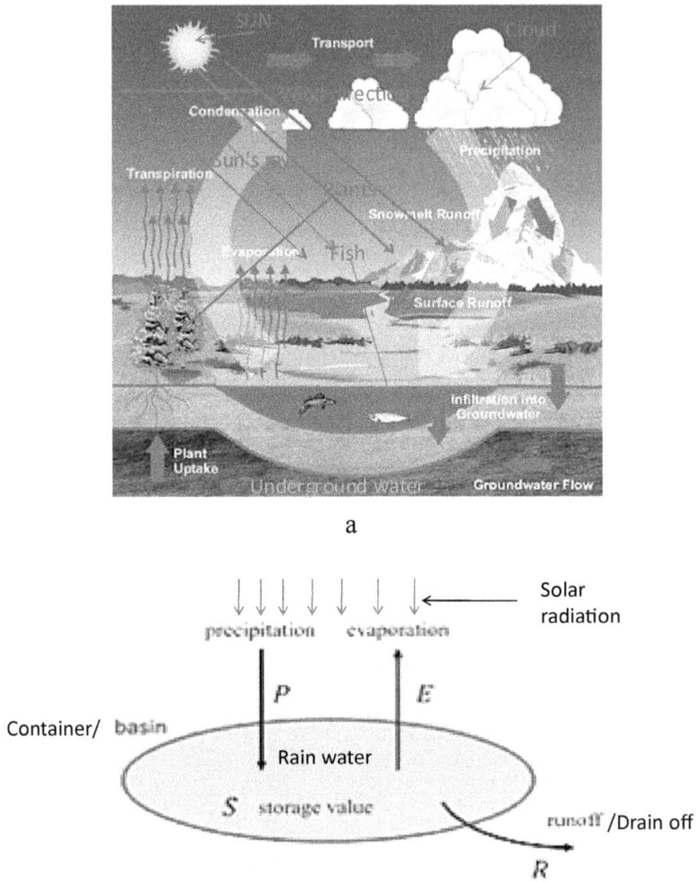

Fig. 2.1 **a** The hydrologic cycle process from water surface of the ground. **b** Water balance system

$$\text{Change in storage} = \text{Inflow (precipitation/rain fall)} - \text{Outflow} \tag{2.3a}$$

General water budget equation in hydrology based on water conservation law for time internal Δt is given by

$$\Delta S = PV - (R + G + E + T) \tag{2.3b}$$

where,

- ΔS Change in water storage capacity
- PV Volume of precipitation
- R Surface runoff
- G Net groundwater flows/infiltration out of the catchment

E Evaporation, (Example 2.1)
T Transpiration (irrigation)
ΔS Change in storage.

Example 2.3 A storage (S) contains 1.55×10^4 m^3 water. If inflow and outflow (run-off) due to precipitation and irrigation are 14.2 m^3/s and 10.6 m^3/s respectively, then find storage (S) reach after an interval of 1 h.

Solution

Initial Storage of rain precipitation $= 1.55 \times 10^4$ m^3

Inflow rate of rain precipitation = 14.2 m^3/s

Inflow volume in one hour $= 14.2 \times 60 \times 60 = 51{,}120$ m^3

Outflow rate for irrigation $(R + G + E + T) = 10.6$ m^3/s

Outflow volume in one hour $= 10.6 \times 60 \times 60 = 38{,}160$ m^3

From Eq. 2.3b, one has

Inflow–Outflow = Change in storage

Table 2.2 Properties of water (saturated liquid)

Temperature		C_p (kJ/kg°C)	ρ (kg/m^3)	μ_k (kg/m s)	K (W/mK)	Pr
°F	°C					
32	0.00	4.225	999.8	1.79×10^3	0.566	13.25
40	4.44	4.208	999.8	1.55	0.575	11.35
50	10.00	4.195	999.2	1.31	0.585	9.40
60	15.56	4.186	998.6	1.12	0.595	7.88
70	21.11	4.179	997.4	9.8×10^4	0.604	6.78
80	26.67	4.179	995.8	8.6	0.614	5.85
90	32.22	4.174	994.9	7.65	0.623	5.12
100	37.78	4.174	993.0	6.82	0.630	4.53
110	43.33	4.174	990.6	6.16	0.637	4.04
120	48.89	4.174	988.8	5.62	0.644	3.64
130	54.44	4.179	985.7	5.13	0.649	3.30
140	60.00	4.179	983.3	4.71	0.654	3.01
150	65.55	4.183	980.3	4.3	0.659	2.73
160	71.11	4.186	977.3	4.01	0.665	2.53
170	76.67	4.191	973.7	3.72	0.668	2.33
180	82.22	4.195	970.2	3.47	0.673	2.16
190	87.78	4.199	966.7	3.27	0.675	2.03
200	93.33	4.204	963.2	3.06	0.678	1.90
210	104.40	4.216	955.1	2.67	0.684	1.66

ΔS = Final storage-Initial Storage = Inflow–Outflow.

or,

Final storage = Initial Storage + Inflow–Outflow = 15,500 + 51,120 − 38,160 = 28,460 m^3

Storage (S) reach after an interval of 1 h = 28,460 m^3

2.3 Carbon Cycle

Carbon (C) cycle as shown in Fig. 2.2a, is the process where carbon compounds are interchanged among the (i) Biosphere: The region between the earth's surface and atmosphere for survival of plants and animals, (ii) Geosphere: It is lithosphere. Figure 1.5, (iii) Pedosphere: It is the outermost layer of the Earth which is composed of soil and living organism (hydrosphere) and (iv) atmosphere of the Earth, Fig. 1.5.

Referring to Fig. 2.2a, the steps involved in the carbon cycle [3] processes are as follows:

(i) Carbon present in the atmosphere in the form of carbon dioxide (CO_2) is absorbed by hydrogen (H_2) isolated by photons after striking the water in plants for photosynthesis, Eq. 1.4.
(ii) The leaves of plants/agricultural products produced due to photosynthesis are then consumed by animals and humans, and carbon is bioaccumulated into their bodies.
(iii) Finally, animals, humans and plants die, and their bodies decompose. After decomposition, the carbon is released back into the atmosphere.
(iv) Some of the carbon which is not released back into the atmosphere becomes fossil fuels to be used as fertilizer and for other applications by humans. During various human activities, carbon returns to the atmosphere.
(v) Carbon formation is very slow process (Fig. 2.2b).

2.3.1 Sources of Carbon Formation

Carbon is an integral part of every life/living organism on planet earth. As shown in Fig. 2.2, the following are the main sources of carbon in the atmosphere [3]:

(a) **Carbon Cycle on Land**

The process of photosynthesis, Eq. 1.4, involves the absorption of CO_2 by hydrogen in plants to produce carbohydrates (biomass). The carbon in the troposphere, Fig. 1.5 is present in the form of carbon dioxide (CO_2) and carbon present in troposphere of the atmosphere is due to natural processes such as respiration from the plant, burning

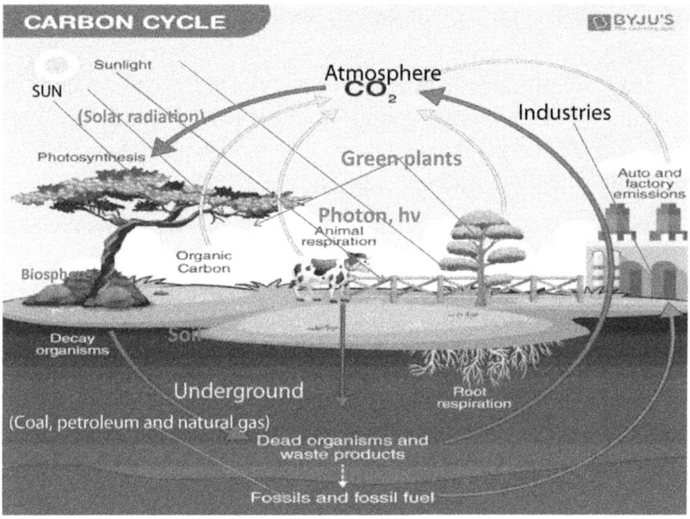

a

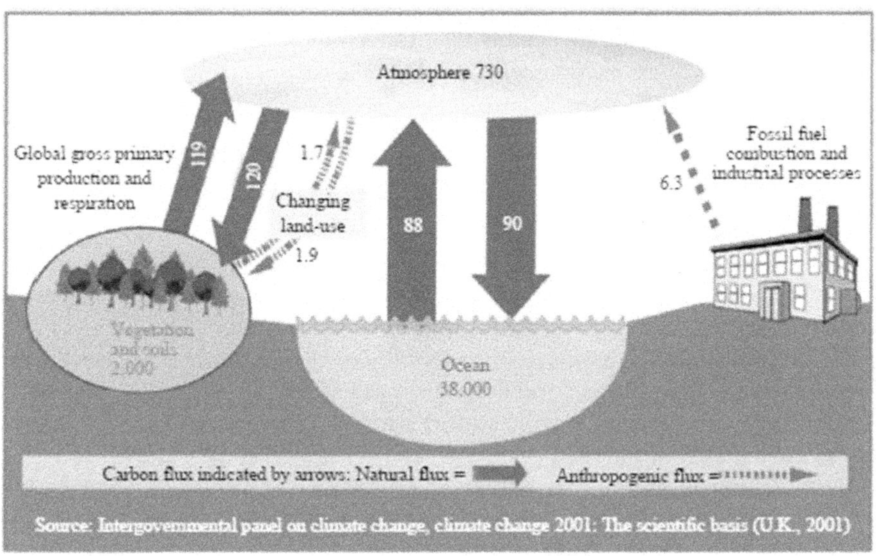

b

Fig. 2.2 **a** Carbon (C) cycle formed by various processes [3]. **b** Carbon (C) cycle formed by various processes (all numbers in billion metric tons) [4]

fossil fuel (biomass) and coal/petroleum/natural gas-based industrial installation for various applications.

Further, the majority of the carbon exists in body of living organisms (Biotic) in the form of carbon dioxide through respiration. The role of decomposers in the body of biotics is to eat the dead organism and return the carbon from their body back into the atmosphere. This chemical process is given by

$$(CH_2O)n + O_2 \rightarrow CO_2 + H_2O \tag{2.4}$$

(b) **Carbon cycle due to oceanic**

On the basis of ecologically, oceans absorb more carbon (C) than it releases to atmosphere and hence finally it is known as "**carbon sink**."

Marine animals (living organisms/biotics), namely birds, fishes, reptiles, sea lions, seals, sharks and rays, squid and octopuses, whales and dolphins etc. convert carbon to calcium carbonate ($CaCO_3$). After the death of living organisms with calcium carbonate shells, their dead bodies decompose, leaving behind their hard shells. These hard shells accumulate on the seafloor and are broken down by the sea waves and compacted under enormous pressure, forming limestone. When these limestone rocks are exposed to the environment (air), then the carbon is released back into the atmosphere as carbon dioxide.

2.3.2 Importance of Carbon Cycle

The presence of carbon dioxide in atmosphere is found to be small traces (carbon dioxide $\approx$ 0.04%) but it plays a vital role in balancing the energy and traps the long-wave radiations from the sun to protect all living organism on the planet earth. Therefore, it acts like a blanket over the planet earth. If the carbon cycle in atmosphere is disturbed by any means, it will result in serious consequences such as climatic changes and global warming.

Furthermore, all known living organisms (biotics) on earth are based on carbon and hence, the carbon cycle, along with the nitrogen cycle and oxygen cycle, play a vital role in the existence of life on earth.

2.3.3 Emission and Absorption of Carbon (C)

As explained in Chap. 1, the short wavelength solar radiation after reflection from ground surface is absorbed by ground surface and finally, it is converted into long wavelength radiation which is not allowed to escape into extra-terrestrial region and hence it is stored in the earth's climatic system (atmosphere). Further, it has been seen that that the oceans, earth and atmosphere are the storehouse of carbon dioxide of (thirty-eight thousand), 2,000 (two thousand) and 730 (seven hundred and thirty) billion metric tons respectively, Fig. 2.2b. The annual oceanic store of CO_2 control the movement of carbon dioxide between ocean and the atmosphere (88 billion metric tons emission and 90 billion metric tons emission: the net annual absorption of 2 billion metric tons by natural flux). Similarly, the net annual absorption of CO_2 by vegetation and soil is (120–119) = 1 billion metric ton. **So, the total net annual absorption of CO_2 by natural flux becomes three (3) billion metric tons**. Now by the anthropogenic flux, the net annual absorption of CO_2 by vegetation/soil is (1.9–1.7) = 0.2 billion metric ton and 6.3 billion metric ton iemission from power plants. **Hence, the net total annual emission of CO_2 is (6.3–0.2) = 6.1 billion metric tons**. Finally, it will be observed that the anthropogenic activities contribute a net value of about 6.1 billion metric tons of carbon and the natural flux removes a net value of about 3 billion metric tons of carbon from the atmosphere. The whole cycle injects annual 3.1 billion metric tons of carbon in the atmosphere which increases the level of CO_2 in atmosphere by industrialization to meet the need of human **[one billion metric tons of carbon = 1** $\times 10^{15}$**g]**.

Further, the oceans occupy more than 70% of the earth's surface, and hence are treated as a large reservoir of CO_2. But the presence of CO_2 (carbon dioxide) in oceans is a function of ocean water temperature due to solubility of CO_2 (carbon dioxide) gasses depending on the temperature. Carbon dioxide (CO_2) from the oceans is released into the atmosphere for higher global ambient air temperature with respect to ocean temperature, especially in summer, and vice-versa (diffusion of CO_2 from atmosphere to ocean) in winter for lower global ambient air temperature. Global ambient air temperatures depends on the Sun's output due to movement of the Earth around the Sun's elliptical orbit. Figure 2.4 shows that one will easily notice that the natural flux is much more than the anthropogenic flux.

Tables 2.3 and 2.4 provide the information about different carbon reservoirs in terrestrial region and direction of carbon flux by between different reservoirs. Figure 2.2b and Tables 2.3 and 2.4 has similarity on both carbon reservoir and carbon fluxes. The average time that carbon atoms spend in a reservoir is called the **residence time**. It can be defined as follows:

$$\text{Residence Time (ET)} = \frac{\text{Total carbon in reservoir in billion matric tones}}{\text{Total flux from resevoir in billion matric tones/year}} \tag{2.5}$$

Table 2.3 Different carbon reservoirs in terrestrial region [5]

S.No	Carbon reservoir	Billion metric tons of carbon
1	Ocean surface	1,000
2	Ocean life	6
3	Organic material in ocean	1,000
4	Deep ocean water	38,000
5	Ocean sediments	3,000
6	Sedimentary rocks	100,000,000
7	Soil	1,600
8	Fossil fuels	4,000
9	Living land organisms	600
10	Atmosphere	750
11	Total carbon reservoir	100,049,956

Table 2.4 Direction of carbon flux between different reservoirs [5]

S.No	Direction of carbon fluxes movement	Billion metric tons of carbon/year
1	Ocean to atmosphere, F_{sa}	102
2	Atmosphere to ocean	105
3	Ocean surface to deep waters	39
4	Deep waters to ocean surface	37
5	Ocean surface to ocean life	28
6	Ocean life to ocean surface	29
7	Soil to atmosphere	60
8	Life on land to soil	60
9	Life on land to atmosphere	50
10	Atmosphere to life on land	110
11	Deforestation to atmosphere	16
12	Fossil fuel combustion to atmosphere	5.4

Example 2.4 Evaluate the residence time of carbon in the atmosphere.

Solution

From Tables 2.3 and 2.4 one has:

Total amount of carbon in the atmosphere reservoir = 750 billion metric tones (Gega tones, Gt), Table 2.3 and

The total flux out (Atmosphere to ocean, 105 Gt + Atmosphere to life on land, 110 Gt). = 215 Gt, Table 2.4

From Eq. 2.5, one has:

$$\text{Residence Time (ET)} = \frac{\text{Total carbon in reservoir atmosphere in billion matric tons}}{\text{Total flux from resevoir in atmosphere in billion matric tones/year}}$$
$$= \frac{750}{215} = 3.5 \text{ years}$$

Referring to Fig. 2.2b, the following equations gives a relation between the carbons fluxes movement, F's [units of pentagrams per year (One petagram, Pg is 10^{15} g)] and the carbons masses reservoir, M's [in units of petagrams, Pg], [6]. It is important to mention here that flux (anthropogenic flux) from burning fossil fuel to atmosphere, F_{fa} is an input to atmosphere, yet to be specified due to fast variation amount of fossil fuel for power generation.

$$\text{Flux from atmosphere to sea surface } (F_{as}) = 0.143 M_a \tag{2.6a}$$

$$\text{Flux from sea surface to atmosphere } (F_{sa}) = 10^{-25} (M_s)^9 \tag{2.6b}$$

$$\text{Flux from atmosphere to terrestrial system (land) } (F_{at}) = 16, 2 \times (M_a)^{0.2} \tag{2.6c}$$

$$\text{Flux from terrestrial system (land) to atmosphere } (F_{ta}) = 0.02 \times M_t \tag{2.6d}$$

$$\text{Flux from deep ocean to surface layer } (F_{ds}) = 0.00129 \times M_d \tag{2.6e}$$

$$\text{Flux from surface ocean to deep ocean layer } (F_{sd}) = 0.45 \times M_s \tag{2.6f}$$

Exercise 2.5 Determine the ocean surface reservoir by using Eq. 2.6b and carbon flux of Table 2.4 (S. No. 1).

Solution
From Table 2.4 (S. No. 1), one has

$$F_{sa} = 102 \text{ billion matric tones per year}$$

From Eq. 2.6b, one has

$$F_{sa} = 10^{-25} (M_s)^9 \text{ or } M_s = \left[\frac{F_{sa}}{10^{-25}}\right]^{\frac{1}{9}} = \left[102 \times 10^{25}\right]^{1/9}$$
$$= \left[1.02 \times 10^{27}\right]^{1/9} \approx 1000 \text{ Billion matric tones}$$

which is accordance of the numerical value of Table 2.3 (S. No. 1).

Exercise 2.6 Determine the soil reservoir by using Eq. 2.6d and soil flux of Table 2.4 (S. No. 7) and Fig. 2.2b

Solution

From Fig. 2.2b, the carbon flux from vegetable and soil is 120 billion metric tons of carbon/year. From Table 2.2b, carbon flux from soil is 60 billion metric tons of carbon/year and life land to atmosphere is 50 billion metric tons of carbon/year and hence total carbon flux from soil and life land becomes 110 billion metric tons of carbon/year which is approximately equal to 120 billion metric tons of carbon/year. For calculation of carbon reservoir from soil/terrestrial land ($\boldsymbol{F}_{\mathbf{ta}}$), we will only consider carbon flux from only soil is 60 billion metric tons of carbon/year.

By using Eq. 2.6d, we get

Flux from terrestrial system (land) to atmosphere (F_{ta}) $= 0.02 \times M_t$

or,

$M_t = \frac{F_{\text{ta}}}{0.02} = \frac{60}{0.02} = 3{,}000$ billion metric tons of carbon.

The above value is about double of 1600 billion metric tons of carbon (S. No. 7) of Table 2.3. However, there is similarity in order between carbon reservoir between Fig. 2.2b and Table 2.3 because [2000, Fig. 2.2b – 600, Table 2.3 (S. No. 9)] = 1200 billion metric tons of carbon.

2.4 Nitrogen Cycle

The majority of atmospheric gas is about 78% [7] which is the largest source of nitrogen (N). The **nitrogen cycle** is a biogeochemical cycle. Biogeochemical cycles generally refer to the movement of nutrients and other elements between biotic (living organism) and abiotic (non-living organism). The living organism and nonliving organisms are (a) human beings, bacteria, animals, birds, fungi, plants, fish, protoctists (nonbacterial organisms), micro-organism etc. and (b) rock, sunlight/solar energy, weather, climate, soil, air, moisture, minerals, natural events such as rock falls or earthquakes, stones, paper, electronic goods, books, buildings, automobiles and more receptively. The nitrogen (N) is converted into multiple chemical forms such as organic nitrogen, ammonium (NH^{+4}), nitrite (NO^{-2}), nitrate (NO^{-3}), nitrous oxide (N_2O), nitric oxide (NO) or inorganic nitrogen gas (N_2) [6]. It circulates around the atmospheric, terrestrial region (Fig. 2.3), and marine ecosystems (Fig. 2.4). The conversion of nitrogen (N) can be carried out through both biological and physical processes which includes fixation, ammonification, nitrification, and denitrification. However, atmospheric nitrogen has limited availability for biological processes which leads to a scarcity of usable nitrogen in many types of ecosystems (Fig. 2.3).

Ecologists need the knowledge of nitrogen cycle. Its availability in atmosphere can affect the rate of ecosystem processes namely primary production of living organism and decomposition after death. Human activities such as (i) fossil fuel combustion, (ii) use of artificial nitrogen fertilizers, and (iii) release of nitrogen in wastewater have dramatically altered the global nitrogen cycle [8–11]. Human modification of the global nitrogen cycle can negatively affect the natural environment system (living organism) along with human health [12, 13]

It is also necessary to know about basic idea about bacteria and its role.

2.4.1 Various Processes of Nitrogen (N) Cycle

Referring to Fig. 2.3, the following are nitrogen processes:

(a) **Nitrogen fixation**

The conversion of nitrogen gas (N_2) from atmosphere with nitrogen fixing bacteria into nitrates (NO_3^-) and nitrites (NO_2^-) through atmospheric, industrial and biological processes is known as **nitrogen fixation**. Nitrogen-fixing bacteria include (a) Rhizobium: it is associated with plants in the pea family, and (b) Azospirillum species: These are associated with cereal grasses. Between 5 and 10 billion kg per year of N_2 are fixed by lightning strikes. However, most fixation is done by free-living N_2 fixing bacteria (Azotobacter). It is aerobic processes which needs oxygen for respiration. These bacteria have the nitrogenous enzyme that combines gaseous nitrogen (N_2) with hydrogen to produce ammonia (NH_4).

The process of Nitrification reaction is given by

$$2NH_3 + 3O_2 \rightarrow 2NO_2^- + 2H^+ + 2H_2O$$

$$2NO_2^- + O_2 \rightarrow 2NO_3^- \tag{2.7}$$

(b) **Assimilation**

Nitrogen assimilation is the process by which plants take in the nitrogen compounds from the soil [ammonium ammonia (NH_4), nitrite ions (NO_2^-), nitrate ions (NO_3^-), Fig. 2.3 and Eq. 2.7] with the help of their roots. It is used in the formation of the plant and animal proteins. Plants and other organisms, which cannot utilize nitrogen molecules directly, depend on the absorption of nitrogen as nitrates or ammonia. This is the way in which it enters the food web for the primary consumers (animals/humans etc.) to eat the green plants.

(a) **Ammonification**

After death of living organism namely the plants, animals, fish and human etc., if they are buried in the ground, then the process of decomposition take place. These processes produce ammonia (NH_4), which is further utilized for other biological processes. The nitrogen present in the organic matter (biological material in the process of decaying or decomposing) is released back into the soil. It is referred as

Anaerobic ammonia oxidation. The decomposers, namely bacteria or fungi present in the soil, convert the organic matter back into ammonium (Fig. 2.3).

Example 2.5 Find out balanced formula for anammox (Anaerobic ammonia oxidation) by using ammonia (NH_4) produced after decomposition of dead body with nitrite (NO^{-2}).

Solution

The following is the balanced formula

$$NH^{+4} + NO^{-2} \rightarrow N_2 + 2\,H_2O\,(\Delta G^\circ = -357\ \text{kJ mol}^{-1})$$

During this process amount of heat releases is -357 kJ mol^{-1} = $-(357 \times 0.278)$ = -99.246 Wh mol^{-1} (1 mol = 6.02×10^{23}molecules}.

(b) **Denitrification**

Denitrification is the process for converting nitrate (NO_3^-) as an electron acceptor in the place of oxygen during respiration into gaseous nitrogen (N_2) which is finally sends back into the atmosphere. This is the reduction of nitrates (NO_3^-) back into

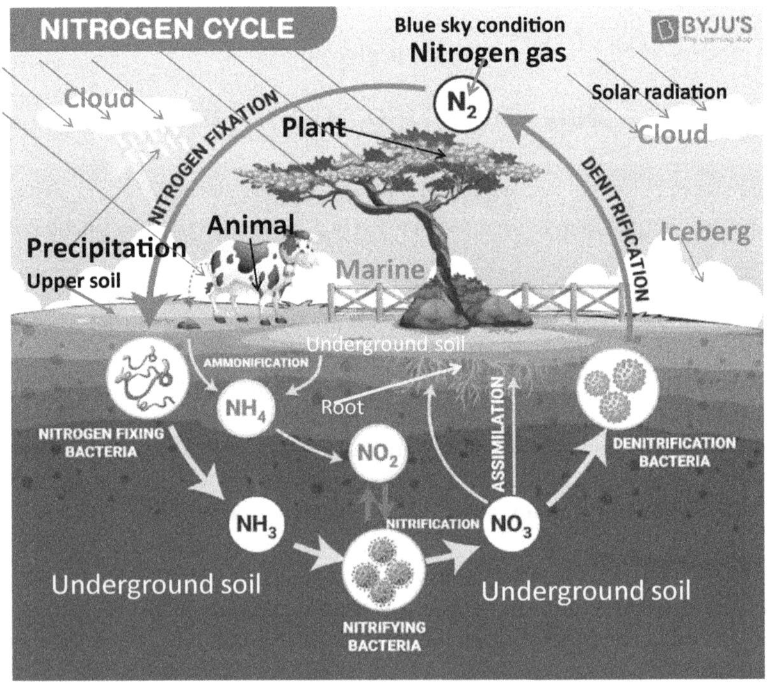

Fig. 2.3 Nitrogen (N) cycle formed by various processes [6]

nitrogen gas (N_2) in the atmosphere, and is referred as nitrogen cycle. This process of the Nitrogen Cycle. This process is the final stage and occurs in the absence of oxygen under anaerobic conditions. This process is performed by bacterial species such as Pseudomonas and Paracoccus, Denitrification generally occurs in free-living microorganisms as well as obligate symbionts of anaerobic ciliates [14].

(c) **Nitrogen cycle in marine/ocean ecosystem**

The process of the nitrogen cycle in the marine/ocean ecosystem occurs in the same manner as in the terrestrial ecosystem as explained earlier. In this case, the nitrogen-containing compounds (N_2) in the atmosphere fall into the marine/ocean through the precipitation/runoff (Fig. 2.1). There are different players [15] and modes of transfer for nitrogen (N) in the ocean. It is carried out by marine bacteria unlike in the terrestrial ecosystem. The nitrogen in the marine/ocean as sediments (material which is broken down by processes of weathering and erosion, and subsequently transported by the action of wind, water, or ice or by the force of gravity) gets compressed over long periods and forms sedimentary rock, and further these sedimentary rocks move to land. The available nitrogen (N) from these rocks is released into plants due to the weathering of the rocks. The nitrogen (N_2) cycle in the marine/ocean environment is shown in Fig. 2.4.

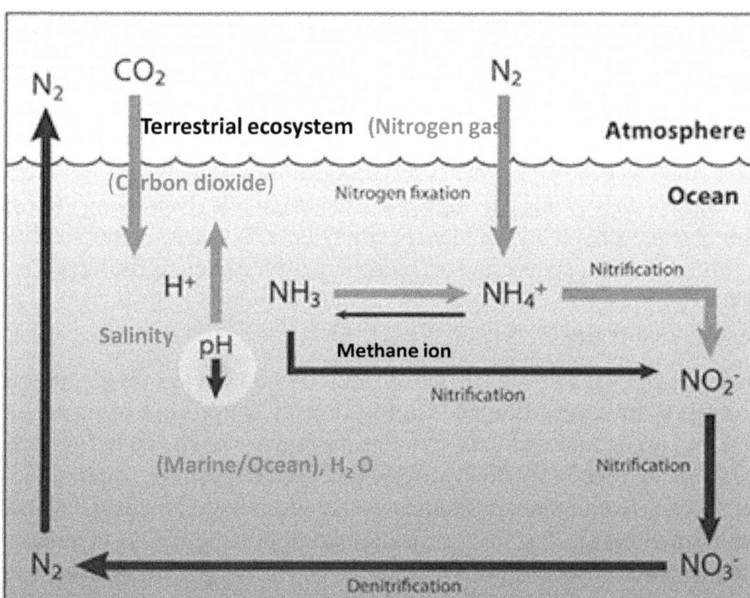

Fig. 2.4 Nitrogen (N_2) cycle in the marine/ocean environment [17]

2.4.2 Importance of Nitrogen Cycle

The nitrogen cycle has the following advantages:

(a) It helps the plants to synthesize chlorophyll from the nitrogen compounds (NH_4, NO_2^-, NO_3^-), through root media of plants.
(b) It also helps in converting inert nitrogen gas (N_2) into a usable form for the plants through the biochemical process.
(c) During ammonification, the active bacteria help to decompose the dead animal and plant matter. This helps to clean up the environment.
(d) Nitrates (NO_3^-) and nitrites (NO_2^-) are released into the soil. This helps in enriching the soil with the necessary nutrients required for cultivation of agricultural product.

Humans are also responsible for creating nitrogen and its compounds in the atmosphere from their activities, such as the combustion of fuels and the use of nitrogen fertilizers. Inorganic fertilizers containing nitrogen (N) are washed away in lakes, rivers, and ponds resulting in eutrophication.

2.5 Oxygen (O_2) Cycle

The oxygen (O_2) cycle plays an impotent role along with the carbon (C) cycle (Eq. 2.2, Sect. 2.3) and nitrogen (N) cycle (Eq. 2.3, Sect. 2.4) for the survival of all living organisms (biotic), including wildlife and humans on the planet Earth. Referring to Fig. 2.5 the oxygen (O_2) cycle is a biological/biochemical process required to maintain the oxygen level in all living organism [16] (biotic) by moving through three main components/spheres of the planet earth (Fig. 1.5). These components include (a) atmosphere, (b) lithosphere and (c) biosphere. Lithosphere is the largest reservoir of oxygen.

Following are the stages of the oxygen cycle:

Stage 1: All green plants, which contain about 90–95% water (H_2O), interact with photons (h ν) of solar radiation; see Fig. 1.6 for the process of photosynthesis. Finally, oxygen is released back into the atmosphere as shown in Fig. 2.5.
Stage 2: All aerobic/biotic/living organisms need oxygen for respiration.
Stage 3: Animals/human/wildlife/aquatic exhale carbon dioxide (CO_2) back into the atmosphere (Fig. 2.5). The exhaled carbon dioxide (CO_2) is again utilized by the plants for photosynthesis (see Eq. 1.4) and oxygen is balanced within the atmosphere.

Following are the applications of oxygen (O_2) [16, 18]:

(a) **Breathing**: It is the main application of oxygen (O_2). In this physical process, all living organisms including plants, wildlife/aquatic and humans first take/

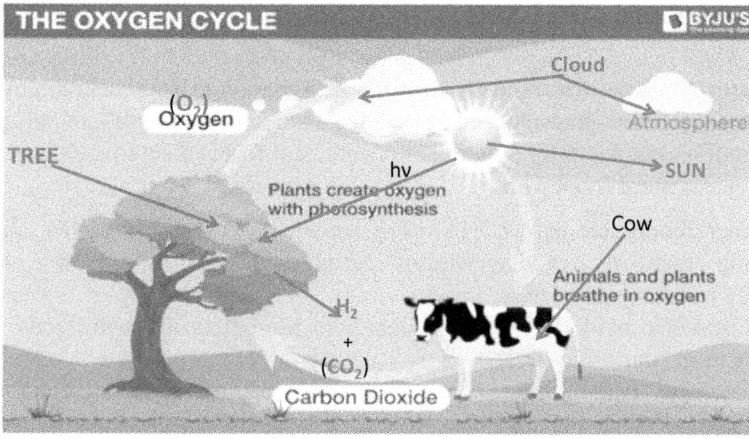

Fig. 2.5 Oxygen (O_2) cycle between plant (tree) and animal/human [14]

inhale oxygen (O_2) from the outside environment into the cells of an organism and then discharge/exhale carbon dioxide (CO_2) back into the atmosphere.

(b) **Decomposition**: After the death of living organisms including plants, wildlife and humans, the natural processes of decomposition takes place in the presence of oxygen (O_2). This is known as the aerobic process. During decomposition, the dead bodies of animals/plants/humans decay with time inside the ground. The organic matter along with the carbon, oxygen, water and other components are returned into the soil and air. This process is carried out with help of fungi, bacteria and insects which are collectively called the decomposers. The whole aerobic process needs oxygen and releases carbon dioxide (CO_2).

(c) **Combustion/burning**: This is also an important process. Oxygen (O_2) is required during combustion/burning of organic materials (fossil fuels, plastics, dead bodies of living organism, and wood etc.) and carbon dioxide is sent back into the atmosphere.

(d) **Rusting**: This process also needs oxygen. During oxidation process over exposed, wet metals like iron or alloy rusting takes place over a period of time. In this case, new compounds of oxides are formed by the combination of oxygen, moisture with the metal.

Following are some properties of oxygen:

i. Molar mass = 15.999u
ii. Density: (a) At STP = 1.429 g/L and (b) When Liquid = 1.141 g/cm^3
iii. Melting Point = −218.79 °C
iv. Boiling Point = −182.962 °C

2.6 Phosphorus (P) Cycle

Phosphorus (P) is an important element for all living organisms including wildlife and humans. It is non-renewable energy source. Humans contain 80% of phosphorus (P) in teeth and bones [16]. It is in solid state, and forms a significant part of the following structural framework.

(a) **Deoxyribonucleic acid** (DNA): The hereditary material in almost all other living organisms including wildlife and humans. Most of cells in a person's body have the same DNA.
(b) **Ribonucleic acid** (RNA): A nucleic acid present in all living cells. It has structural similarities to DNA.
(c) **Adenosine triphosphate** (ATP): An energy-carrying molecule. It is found in the cells of all living organism (biotic). It captures chemical energy obtained from the breakdown of food molecules and releases it to fuel other cellular processes.

The phosphorus cycle is a very slow process. Rain washes phosphorus from rock into soil. It is absorbed in the soil by the organic matter which is later used for various biological processes. There is no phosphorus in the atmosphere like oxygen (O_2), nitrogen (N_2) and Carbon (C).

The phosphorus (P) cycle has following important steps:

(a) **Weathering**: Phosphorus is found in rocks in abundance and hence the phosphorus cycle starts in the Earth's crust. The phosphate salts are broken down from rocks by weather (rain) and then washed away and mixed into the soil. However, the presence of phosphorus (P) in the soil is very low. For that reason, farmers use external phosphate fertilizers on agricultural land for higher yield.
(b) **Absorption by Plants**: Phosphate salts are also dissolved in water which is absorbed by the plants.
(c) **The aquatic plants**: It absorbs inorganic phosphorus from lower layers of water bodies. Growth of aquatic plants is affected if phosphate salts do not dissolve in water properly.
(d) **Absorption by Animals**: The animals absorb phosphorus either from the plants or by consuming it as feed. The rate of the phosphorus cycle is faster in plants and animals in comparison to rocks.
(e) **Return to the Environment through Decomposition**: After death of the plants and wildlife, human, the decomposition of body take place with microorganisms. Decomposition of wildlife and human body is faster than plants. During this process, the organic form of phosphorus is converted into the inorganic form. Further, it is used/recycled to soil and water for increasing productivity of agricultural product.

Soil and water will end up in sediments and rocks. Due to weathering, phosphorus will again be released, thus, the phosphorus cycle is complete.

The use of fertilizers by human activities such as fertilizer in agricultural sector has a great impact on the phosphorus cycle. Overuse of these fertilizers also reduces the fertility of the soil and becomes harmful to the microorganisms present. After overuse, the phosphorus is washed away to nearby water by weathering, then aquatic life is badly affected.

2.6.1 Case Studies for Phosphorous

In order to apply fertilizer for agricultural purposes, one should have knowledge of the nutrient content of soil, along with soil type, depth, current pasture or crop type and history. The unit of phosphorous is I unit P (1 kg P). The basic component of fertilizer is in ratio as nitrogen (N): phosphorous (P): potassium (K): sulfur (S) which is generally written on bags/packets of fertilizer. For example:

(a) **Single superphosphate fertilizer**: The constituents of nutrients are 0:8.8:0:11 which means in 100 kg of fertilizer, there are no nitrogen, 8.8 kg phosphorus, no potassium and 11 kg sulfur.
(b) **Starter 15 fertilizer**: It is in ratio of 14:12.7:0:11 which means in 100 kg of fertilizer, there is 14 kg nitrogen, 12.7 kg P, no potassium and 11 kg sulfur.

Following expression can be used to calculate amount of fertilizer as

$$\text{Amount of fertilizer in kg per ha} = \frac{\text{kg per ha nutrient}}{\%\ \text{nutrient in fertilizer}} \times 100 \tag{2.8}$$

Example 2.6 Calculate amount of fertilizer for case of single superphosphate fertilizer by using 20 units (kg)/ha of phosphorus (P) [nutrient].

Solution

For single superphosphate fertilizer, there is 8.8% P (known)

From Eq. 2.8, one gets

Amount of fertilizer in kg per ha $= \frac{20\,\text{kg/haP}}{8.8\text{P}} \times 100 = 227$

Example 2.7 Calculate amount of P nutrient in kg per ha for 125 kg per ha of fertilizer for single superphosphate fertilizer.

Solution

From Eq. 2.8, one can rewrite

$$\text{kg per ha nutrient} = \frac{\text{Amount of fertilizer in kg per ha} \times \%\ \text{nutrient in fertilizer}}{100}$$

$$= \frac{125 \times 8.8}{100} = 11\ \text{kg/ha}$$

2.7 Wind Cycle

The non-uniform heating of surface of earth and sea by solar radiation coming from the sun is responsible for creation of wind. It is due to fact that solar radiation in the terrestrial region which reaches the Earth during daytime comes at different rates (W/m^2) and is absorbed by land, plants and water. After absorption by the Earth's surface, it is heated, and the thermal energy associated with it is transferred by convection to the air above it. Further, air is heated with higher pressure and moves upward towards lower pressure (temperature) due to lower density of air. The heavier cooled air from the top rushes in to take the space created by moving hot air moving upward, thus wind is created. Refer to Fig. 1.5. The atmosphere is comprised of four layers, namely troposphere, stratosphere, mesosphere, and thermosphere, which are stationary and more than 100 km above the crust of the Earth. In this region, there are many balanced gases flowing around, and each of them serves a purpose. The wind is the result of differences in pressure in the atmosphere and flows between cooled air masses. It moves from high-pressure above ground to low-pressure atmospheric regions as mentioned above. The wind always moves from east to west due to the rotation of the Earth. At night, (off-sunshine hour), the winds are reversed because the air over land is cooler than over water. The **day wind cycle** can be understood in four steps as follows:

1. Sun heats the air over surface of land.
2. Hot air rises from ground surface towards cooled troposphere.
3. Cooled air from troposphere moves into fill the space created by hot air above the ground surface.
4. This moving air is the formation of wind cycle.

The day wind cycle has been shown in Fig. 2.6.

The energy associated with wind energy created by solar energy as mentioned above is in the form of kinetic energy. This kinetic energy can be converted into electricity by using a wind turbine of cross-sectional area of A. The details about this will be discussed in Chap. 3. The wind energy is renewable because it depends on solar energy. The wind turbines first convert the kinetic energy present in the wind to mechanical power and then into electricity.

The formula for kinetic energy associated with wind is written as

$$\text{Wind power}(P) = \frac{1}{2}\rho A(u_0)^3 = \frac{1}{2}\rho\pi R^2(u_0)^3 \tag{2.9}$$

where, $\rho = 1.23$ kg/m^3 is density of moving air, $A = \pi R^2$is swept area of blades, R is the radius of the blades and V is the wind velocity.

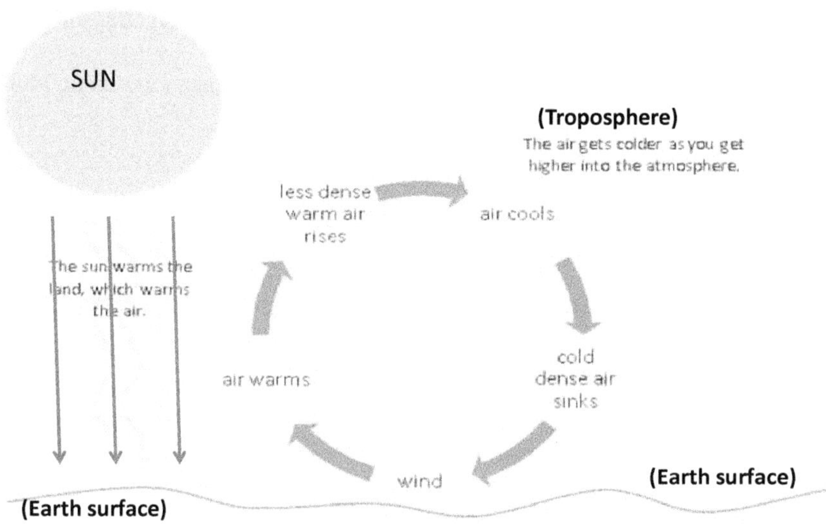

Fig. 2.6 Wind cycle formation

Example 2.8 Find out the power associated with the wind for the wind speed of 10 m/s for a given turbine blade length of 25 m.

Solution

Substitute the given values in Eq. 2.9, one gets

Wind power $(P) = \frac{1}{2} \times 1.2 \times 3.14 \times (25)^2 \times (10)^3 = 1,177,500\,\text{W} = 1,177.5\,\text{kW}$
$\text{P} = 1,177.5\,\text{kW}$

2.8 Bacteria

As discussed in Chap. 1 (Sect. 1.4), algae was the first living organism in the form of bacteria formed due to interaction between photons of solar radiation and melted water from the sea due to the global greenhouse effect. Hence, bacteria are considered to be one of the oldest forms of life on planet Earth. Bacteria have been created around for a *very* long time ago (3.5 billion years) due to the greenhouse effect after the formation of the Sun and the Earth. The earliest most common fossils are of **prokaryotes** [21–23] which were single cells, without a nucleus as seen in blue-green algae, bacteria and mycoplasma. They multiplied fast unlike eukaryotic cells. Their size ranged from 0.2 to 10 μm, which is about 10 times smaller than most plant and animal cells. Most of them are only about 1–2 μm in diameter. On the other hand, eukaryotic cells are found in plants, animals, fungi, and protists and human has nucleus whose bodies consists of million/trillions of cells. Bacteria also contribute to a range of essential soil processes involved in the cycling of carbon (C), nitrogen (N) and phosphorus (P).

There are both useful and harmful bacteria for living organism including plants/wildlife/human which are as follows.

2.8.1 Useful Bacteria

The useful/beneficial bacteria in different forms with benefits are listed below: Convert milk into curd—Lactobacillus or lactic acid bacteria

(a) Ferment food products (cultured milk and yogurt wine beer cider tempeh miso kimchi sauerkraut, sausages, pickles etc.): Fermentation bacteria are anaerobic which acts in absence of oxygen (O_2). It uses organic molecules as their final electron acceptor to produce ferment food product. Bacillus species produce a huge variety of antimicrobial peptides (AMPs) i.e. lactic acid. It has their long been used as a food preservative by direct addition of the bacteria to foods.
(b) It helps in digestion and improving the body's immunity.
(c) It is a production of antibiotics. It is used in the treatment and prevention of soil bacterial infections.
(d) Bacteria are the most abundant microbes in the soil. There can be billions of bacteria in a single gram of soil.

2.8.2 Harmful Bacteria

Some bacteria are responsible for many of the **infectious diseases** (multitude of illnesses) like pneumonia, tuberculosis, diphtheria, syphilis, tooth decay. Their effects can be rectified either by taking antibiotics or by consuming appropriate medicinal plant or by using prescribed medication. However, taking precaution is effective. These precautions include application of thermal energy/solar energy/heat, disinfectants, UV radiations, pasteurization, and boiling.

Problems

Problem 2.1 Plot the curve between partial vapor pressure, P(T) with temperature in °C.

Solution See Example 2.1 and Table 2.1.

Problem 2.2 Evaluate rain water collected for roof area of 5 m × 10 m with rain fall of 20 mm.

Hint: Use Eq. 2.2.

Problem 2.3 How much precipitation in mm is needed to collect rain water to feed four family members in one year with a roof area of 10 m × 10 m, if each member consumes 4 L of water daily.

Hint: Use Eq. 2.2 and 1 L = 0.001 m^3

Problem 2.4 Evaluate the residence time of carbon in the ocean.

Hint: Use Example 2.4 and Tables 2.3 and 2.4 respectively.

Problem 2.5 Find the following nutrients in kg per ha for starter 15 fertilizer for 125 kg er ha fertilizer.

Hint: Use Example 2.8 and ration of nutrients as 14:12.7:0:11 for starter 15 fertilizer.

Problem 2.6 Find wind velocity for power to be obtained as $P = 1, 177.5\,kW$

Hint: See Example 2.8

Objective Questions

2.1 First/second collected rain/precipitated water is

(a) Good for heath
(b) Bad for health
(c) No effect on health
(d) All of these

Answer: (b)

2.2 The collected rain/precipitated water after two/three rain is

(a) Good for heath
(b) Bad for health
(c) No effect on health
(d) All of these

Answer: (a)

2.3 Rain/precipitated water contains

(a) All minerals of potable water
(b) No minerals
(c) Impure water
(d) All of these

Answer: (a)

2.4 The hydrological cycle took place

(a) Between the Earth and the atmoshphere
(b) Between the sea and earth through the atmosphere
(c) Between the sea and the atmosphere
(d) In all cases

Answer: (b)

2.5 Plants take nitrogen from

(a) Atmosphere through leaves
(b) Environment through leaves
(c) Soil through roots
(d) None of these

Answer: (c)

2.6 Carbon (C) from the atmosphere is basically absorbed

(a) By plants for photosynthesis
(b) By the Earth from roots of plants
(c) By the sea for fish
(d) All of these

Answer: (a)

2.7 The decamped body of an animal releases

(a) Carbon into the atmosphere indirectly
(b) Carbon into the atmosphere directly
(c) Carbon into the soil
(d) All of these

Answer: (a)

2.8 The nitrogen (N) consumed by humans is absorbed

(a) Directly
(b) Indirectly
(c) Both directly and indirectly
(d) None of these

Answer: (b)

2.9 The carbon (C) consumed by human is absorbed

(a) Directly
(b) Indirectly
(c) Both directly and indirectly
(d) None of these

Answer: (c)

2.10 The oxygen (O_2) consumed by humans is absorbed

(a) Directly
(b) Indirectly
(c) Both directly and indirectly
(d) None of these

Answer: (a)

2.11 The highest carbon reservoir is

(a) Deep ocean water
(b) Fossil fuels
(c) Atmosphere
(d) Sedimentary rocks

Answer: (d)

2.12 Wind power is

(a) Proportional to swept area of wind turbine
(b) Proportional to square of radius of wind turbine
(c) Proportional to square of wind velocity
(d) All of these

Answer: (a) and (b).

2.13 Wind power (P) is

(a) Proportional to swept area of wind turbine
(b) Proportional to square of radius of wind turbine
(c) Proportional to cube of wind velocity
(d) All of them

Answer: (d)

2.14 Wind power (P) is

(a) Inversely proportional to swept area of wind turbine
(b) Proportional to square of radius of wind turbine, $(R)^2$
(c) Proportional to cube of wind velocity, $(u_0)^3$
(d) All of these

Answer: (b) and (c)

References

1. Water Cycle—Process and its Various Stages—Byju's. https://byjus.com, Biology, Biology Article
2. W. Duangthongsuk, S. Wongwises, Exp. Therm. Fluid Sci. **33**, 706 (2009)

3. Definition, Process, Diagram of Carbon Cycle—Byju's. https://byjus.com, Biology, Biology Article
4. C.A. Mgbemene, C.C. Nnaji, C. Nwozor, Industrialization and its Backlash: focus on climate change and its consequences. J. Environ. Sci. Technol. **9**, 301–316 (2016). https://doi.org/10.3923/jest.2016.301.316
5. Carbon budget 750 3.5 215/gt years gt yr. http://maecourses.ucsd.edu
6. Carbon cycle modelling: illustrations of. https://maths.anu.edu.au, cmaprocvol25-enting
7. S.B. Carroll, S.D. Salt, *Ecology for Gardeners* (Timber Press, 2004), p. 93. ISBN 978-0-88192-611-8
8. Nitrogen Cycle Explained—Definition, Stages and Importance https://byjus.com. 'Biology' Biology Article
9. M.M.M. Kuypers, H.K. Marchant, B. Kartal, The microbial nitrogen-cycling network. Nat. Rev. Microbiol. **1**(1), 1–14 (2011). https://doi.org/10.1038/nrmicro.2018.9.hdl:21.11116/0000-0003-B828-1.PMID29398704.S2CID3948918
10. J.N. Galloway et al., Nitrogen cycles: past, present, and future generations. Biogeochemistry **70**(2), 153–226 (2004). https://doi.org/10.1007/s10533-004-0370-0.S2CID98109580
11. S. Reis, M. Bekunda, C.M. Howard, N. Karanja, W. Winiwarter, X. Yan, A. Bleeker, M.A. Sutton, Synthesis and review: tackling the nitrogen management challenge: from global to local scales. Environ. Res. Lett. **11**(12), 120205 (2016)
12. B. Gu, Y. Ge, Y. Ren, B. Xu, W. Luo, H. Jiang, B. Gu, J. Chang,Atmospheric reactive nitrogen in china: sources, recent trends, and damage costs. Environ. Sci. Technol. **46**(17), 9420–9427 (2012). https://doi.org/10.1021/es301446g. ISSN 0013-936X. PMID 22852755
13. H. Kim, K. Lee, D.I. Lim, S.I. Nam, T.W. Kim, J.Y. Yang, Y.H. Ko, K.H. Shin, E. Lee, Widespread anthropogenic nitrogen in northwestern pacific ocean sediment. Environ. Sci. Technol. **51**(11), 6044–6052 (2017). https://doi.org/10.1021/acs.est.6b05316. ISSN: 0013-936X, PMID 28462990
14. J.S. Graf, S. Schorn, K. Kitzinger, S. Ahmerkamp, C. Woehle, B. Huettel, C.J. Schubert, M.M. Kuypers, J. Milucka, Anaerobic endosymbiont generates energy for ciliate host by denitrification. Nature **591**(7850), 445–450 (2021). https://doi.org/10.1038/s41586-021-03297-6. ISSN 0028-0836. PMC 7969357. PMID 33658719
15. O.M. Moulton, M.A. Altabet, J.M. Beman, L.A. Deegan, J. Lloret, M.K. Lyons, J.A. Nelson, C.A. Pfister, Microbial associations with macrobiota in coastal ecosystems: patterns and implications for nitrogen cycling. Front. Ecol. Environ. **14**(4), 200–208 (2016). https://doi.org/10.1002/fee.1262. hdl:1912/8083. ISSN:1540-9295
16. Oxygen Cycle-Process, Uses, Production and Facts. https://byjus.com, oxyg
17. Nitrogen cycle—Wikipedia. https://en.wikipedia.org, 'wiki', Nitrogen_cycle
18. Oxygen cycle—Wikipedia. https://en.wikipedia.org
19. Steps And Importance Of Phosphorus Cycle—Byju's https://byjus.com, phos
20. Cycle-Process, Uses, Production and Facts. https://byjus.com, Biology, Biology Article
21. Forest soil bacteria: diversity, involvement in ecosystem **81**(2). https://journals.asm.org
22. Soil Bacteria and Fungi—NSW I Fact Sheets I soilquality.org.au https://soilquality.org.au, factsheets, soil-bacteria-and
23. Bacteria—Definition, Structure, Diagram, Classification. https://byjus.com, Biology, Biology Article

Chapter 3
Renewable Energy

3.1 Introduction

The meaning of renewable energy is very clear: It should be (i) clean (environmentally friendly), (ii) infinite, (iii) continuously available, and (iv) sustainable; such as **Solar Energy**. It is sometimes referred as either non-conventional or alternative energy sources. It means sources of energy that are alternative to mostly used the non-renewable/conventional/fossil fuel (coal/ petroleum/natural gas) which are non-sustainable sources, and friendly to the environment and climate.

In addition to solar energy, the most popular renewable energy sources are as follows:

(i) **Biomass energy**: Biomass energy is energy created/generated/produced by once-living organisms, preferably from plants such as trees and corn (sugar corn and pole corn), soybeans etc. by using solar energy for photosynthesis. The dried biomass can be burned to create either thermal energy/heat or electricity known as **biopower**. The biomass produced by photosynthesis is always referred by **biomass fuel/bioenergy**.

(ii) **Hydro energy**: It is also referred as water energy. It is stored gravitational potential energy (ρgh : h is height, ρ is density of water and g is gravitational coefficient) at high altitude due to the hydrogen/water cycle, as explained in Chap. 2 due to raining in hilly areas, in a natural dam. In this case, potential energy with stored water is allowed to fall and is converted into kinetic energy with the help of a generator; it is then converted into electrical power.

(iii) **Wind energy**: Wind is also created by absorption of solar energy by Earth's suface and due to temperature differences; coming air is heated and moves upward and is further driven by pressure differences. It contains kinetic energy (KE) due to its movement. By using wind turbines, it is converted into electricity which is a sustainable renewable energy source. It has smaller impact on the environment than the burning of biomass.

G. N. Tiwari, *Energy, Ecology and Environment*,
https://doi.org/10.1007/978-981-99-3997-8_3

(iv) **Geothermal energy**: It is heat derived/obtained from within the earth. Hot water/steam carries the geothermal energy to the Earth's surface by water pumping. It is used for heating and cooling purposes or can be harnessed to generate clean electricity.

All renewable energy are sustainable energy due to solar energy as its input energy.

As discussed in Chap. 1, solar energy coming from the Sun is transmitted from extra-terrestrial region to terrestrial region and then it reaches to the planet Earth, Fig. 1.2. Solar energy is in the form of electromagnetic waves (thermal energy) as well as in the form of rays (photons). It is intermittent in nature in the terrestrial region and continuously available in the extra-terrestrial region, and hence it is used by all satellites for 24-hour power generation for its uses. It is directly absorbed by all living organisms as well as ground surface/ocean/water ponds etc. and after absorption it is used in various form such as vitamin D by all living organisms namely wildlife, aquatic and humans, and thermal energy for their thermal comfort. Further, solar energy in the form of photons is used mainly for (a) photosynthesis by plants to generate biomass energy as a storage of solar energy and (b) solar cell to generate direct current and voltage (DC) which can be used directly for various applications without storage such as in (a) water pumping in the terrestrial region and (b) satellites in the extra-terrestrial region. Further, hydro energy is obtained from stored water ponds at higher altitudes due to stored potential energy from rainfall (hydrogen/water cycle, Chap. 2).

Based on the above brief discussion, solar energy, which is renewable energy, can be broadly classified as follows:

(a) **Renewable energy without storage**: In this case, there is direct use of solar energy such as vitamin D, thermal heating, wind generation in the form of kinetic energy, photosynthesis with plants and generation of DC power from photo-voltaic (PV) systems. In all cases, there is direct use of solar energy, and after use, it can be stored in the body, batteries, as biomass etc. But all are intermittent due to the intermittent nature of solar energy.

(b) **Renewable energy with storage**: In this case, there is indirect use of solar energy such as hydrogen/water cycle/carbon cycle/nitrogen cycle/oxygen cycle etc. (Chap. 2). In all cycle, solar energy plays an important role to achieve all cycle phenomenon. After completing all cycle, solar energy is indirectly stored and it is used as required. For example, water falls due to the hydrogen/water cycle in hilly regions; the water is stored in the form of natural dams which contain the potential energy. The energy stored will depend on the capacity of the dam to be used for power generation. In this case, solar energy is stored indirectly due to rain and then converted into electrical power as required.

In the next section, we will briefly discuss renewable energy with and without storage.

3.2 Vitamin D

The synthesis (the act of combining elements to form something new) of vitamin D is effectively linked to the availability of solar energy as shown in Fig. 3.1. Vitamin D is the sunshine's free vitamin. It has been produced on the Earth for more than a billion years. The synthesis of vitamin D with the Sun is greatly influenced by season, time of day, latitude (ϕ), altitude (α)/elevation of the Sun, environmental air pollution, skin pigmentation, sunscreen use, glass windows and transparent plastic, and aging etc. It is metabolized sequentially in the liver and kidneys. Vitamin D deficiency and living at higher latitudes like hilly regions have increased risks for creating many chronic diseases like autoimmune diseases, cancers, cardiovascular disease, infectious diseases, schizophrenia and type 2 diabetes. Vitamin D can also affect many bodily functions including bone health and autoimmune diseases. Furthermore, consumption of vitamin D also helps calcium absorption in the human body to give strength to bone and skeletal integrity. Deficiency of vitamin D can be avoided by the following means [1, 2]:

(a) **Sensible direct sun exposure**: Also referred to as sunbathing therapy. It is well known that newborn babies should be covered with thin cotton cloth when exposed to the Sun to avoid harsh heating. This helps the baby gain strong bones and teeth by increasing calcium. Sunbathing therapy should not be done in windy conditions. The shorter the shadow, the bigger the gains of required Vitamin D. The duration may vary between 15 and 30 minutes depending on season (summer/winter). Direct exposure of UVB rays having wavelengths between 280 nm (0.28 μm) and 315 nm (0.315 μm), Fig. 3.2 shows visible wave lengths ranging between 300 and 700 nm (0.30–0.70 μm) increasing the synthesis of vitamin D3 in the human body to increase blood pressure [2, 3]. The decrease in sun exposure to minimal times mentioned above may result in physical and

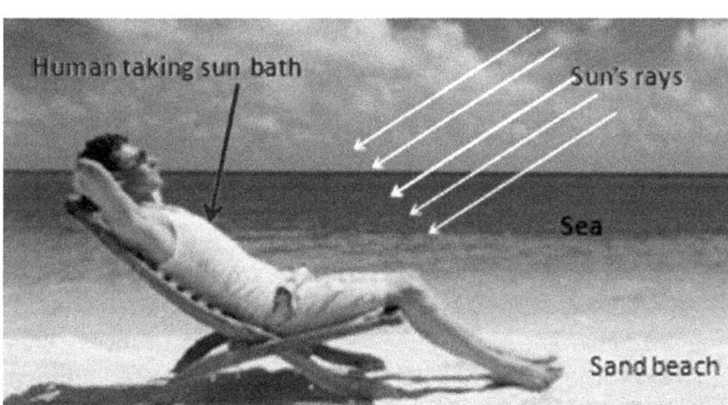

Fig. 3.1 Free sun bath therapy to increase vitamin D in early morning

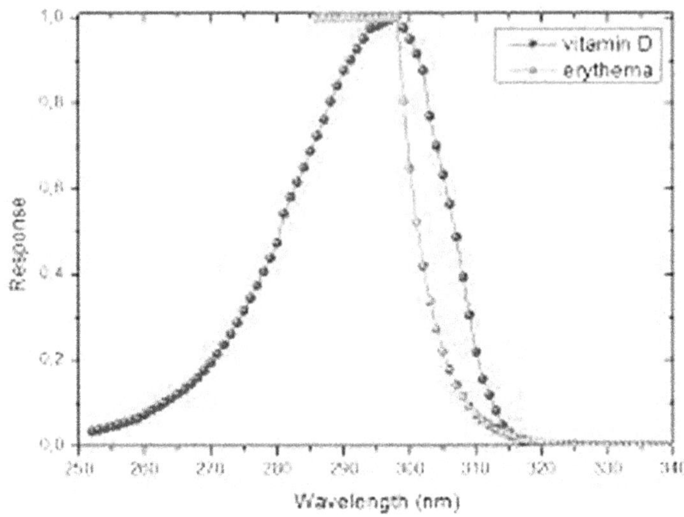

Fig. 3.2 Response of vitamin D action spectrum [2] (blue line) (250–330 nm) and erythema action spectrum (285 and 300 nm.) [3] (red line)

psychological disorders due to deficiency of Vitamin D in the body. Excessive direct exposure to the Sun can harm your skin and lead to skin cancer.

(b) **Consuming more food/fortification with vitamin D**: Following non-vegetarian and vegetarian foods give sufficient vitamin D required by human body:

(A) **Non-vegetarian food**:

(i) **Salmon**: It is a popular fatty fish and a great source of vitamin D.

(ii) **Herring and sardines**: It is a great source of vitamin D and also a fish used as a food around the world as pickled.

(iii) **Canned Tuna**: Due to its flavor, smaller, cheaper and easy to store, it is used by many people. Light canned tuna comes from smaller fish containing smaller amounts of mercury.

(iv) **Egg Yolk**:. The yellow middle part of an egg contains minerals. It is a good source of vitamin D and is a nutritious food.

(B) **Vegetarian food**:

(i) Cod liver oil: It is a popular excellent source of vitamin D supplement. to get nutrients. It also has a history of being used to treat rickets, psoriasis, and tuberculosis.

(ii) **Mushroom**: Many mushrooms are grown in the dark and contain very little D2 which helps raise blood levels of vitamin D. If mushrooms exposed to UV light are absorbed by humans, they can synthesize vitamin D. Some wild mushrooms are excellent sources of vitamin D2 due to their exposure to UV light.

Table 3.1 Vitamin D in microgram (1×10^{-6} g) international unit (IU) requirement by different age groups [4, 5]

S. no.	Age	Daily amount of vitamin D required
1	0–12 months	10 mcg (400 IU)
2	1–18 years	15 mcg (600 IU)
3	19–70 years	15 mcg (600 IU)
4	70 years and above	20 mcg (800 IU)

(iii) **Fortified food as Vitamin D**: Natural sources of vitamin D are limited, for vegetarians. Fortified foods containing vitamin D include milk, soy milk, orange juice, cereals and oatmeal.

(c) **Encouraging ingestion of a vitamin D supplement**: Needed to avoid vitamin D deficiency and associated negative health consequences. The daily requirement of vitamin D for people depends on age, physical condition and any underlying medical conditions. The national institute of health (NIH) provides these general guidelines [4], as given in Table 3.1.

The 15 mcg (600 IU) of vitamin D per day is recommended for pregnant or breastfeeding women after consultation of with a doctor.

3.3 Photosynthesis

As mentioned in Sect. 1.4 about global greenhouse effect which is responsible for creating green algae (micro-organisms) by interaction of photons of solar energy with water of melted ice (water) and oxygen in terrestrial region of planet earth. It is named photosynthesis. In this case, oxygen (O_2) is released as a byproduct, and light ($h\nu$) is considered a major factor to complete the process of photosynthesis; see Eq. 3.1. Further, photosynthesis occurs if light ($h\nu$) interact with plants to convert carbon dioxide and water into glucose/fructose (sugar) and oxygen (Fig. 3.3). Green leaves contain microscopic cellular organelles known as chloroplasts. Each chloroplast contains a green-colored pigment known as chlorophyll. Light energy ($h\nu$) is absorbed by chlorophyll molecules in plant leaves, whereas carbon dioxide and oxygen enter through the tiny pores of stomata located in the epidermis of leaves.

The produced glucose/fructose (Sugar) (see Figs. 1.6 and 3.3) are then sent to the roots, stems, leaves, fruits, flowers and seeds of plants as a energy source to helps them grow. These produce glucose/fructose (sugar) molecules that combine to form more complex carbohydrates like cellulose and starch. Finally, one can say that solar radiation is directly used for photosynthesis. This is referred to light reaction.

$$\mathbf{6CO_2 + 12H_2O + light \rightarrow C_6H_{12}O_6 + 6O_2 + 6H_2O} \tag{3.1}$$

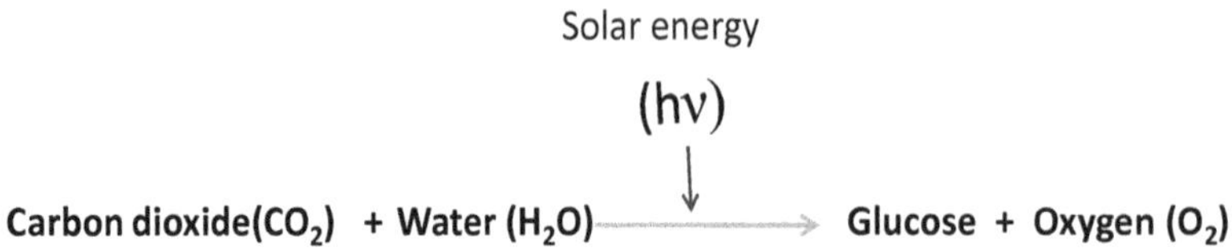

Fig. 3.3 A visual chemical representation of the photosynthesis processes

Main factors required for the photosynthesis process are as follows:

(i) **Light intensity/solar radiation**: It is basic requirement which provides energy ($E = h\,\nu$) from photons. The energy content depends on frequency. For visible wave length ranging between 300 and 700 nm (0.30–0.70 μm), energy content is maximum and hence maximum photosynthesis takes place in this wave length range.

(ii) **The concentration of CO_2**: For higher photosynthesis, one needs higher concentration of carbon dioxide (**CO_2**) in the atmosphere. Generally, carbon dioxide in the atmosphere in the range of 300–400 ppm is enough for photosynthesis. But at more than this concentration, photosynthesis is adversely affects plant growth.

(iii) **Temperature**: For efficient photosynthesis, a optimum temperature range between 25 and 35 °C is recommended. For higher temperatures, the respiration from the plant increases and the water content of plant is reduced, affecting the photosynthesis.

(iv) **Water**: As water is an important factor in photosynthesis as mentioned above and shown in Eq. 3.1. The deficiency of water content in plant scan lead to reduction of photosynthesis due to the resultant reduction in hydrogen production by photons.

(v) **Pollution**: It has been observed that due to increased industrial manufacturing and operation of motor vehicles since World War Two, pollutants and other particulates may settle on plant leaf surfaces This minimizes the absorption of photons and hence reduces the photosynthesis.

The product of photosynthesis is also known as oxygenic photosynthetic bacteria. It is biomass which can be referred to as storage of solar energy as biomass fuel like fossil fuel. Biomass fuel can also be used for generating thermal energy as well as electricity when required, such as when there is no solar energy. Biomass fuel can be transported from one place to another applications such as domestic cooking, housing construction and industry. Many jobs are generated by the biomass industry.

3.4 Photo-Voltaic Module

The state of material is liquid, gas and solid. Further, the solid can be either conducting material or insulating material. There is also another solid material known as semi-conductor. It is made up of silicon, and is an n-and p-type material. There is an excess of electrons and protons in n- and p-type material respectively. If n- and p-type material is joined as p–n junction, electromotive force (emf) is generated due to the flow of electrons from n-type to p-type material, and there is a flow of current from p-type to n-type at the junction. These electrons and protons become neutral at the junction, which is described as a depletion region. The n-type material is more conductive in comparison to p-type material. It is also transparent to photon of solar energy. The thickness of n-type material is less than p-type. If solar energy (photon $= h\nu$) falls on n-type silicon as shown in Fig. 3.4a, it is transmitted to the depletion region and strikes with neutral electrons and holes. Since the energy associated with protons is higher than binding energy between electrons and holes, the electron and hole are separated. After this separation, electrons move toward n-hype material and holes move toward p-type material. Now, electrons cross n-type material due to its lesser thickness and lower conductivity, so electrons move in an external circuit from n-type to p-type material through load and DC current is generated. The direction electron from n-type to p-type and DC current from p- to n-type material has been shown in Fig. 3.4a. This effect is known as **photo-voltaic effect** due to the use of photons. The load may be a bulb, fan, charger etc. Such a unit is known as a **solar cell** because it acts as a cell due to solar energy. The specification of c–Si solar cell under standard test condition, STC (constant solar intensity, $I_0 = 1000\ Wm^2$ and $T_a = 25\ °C$) has been given in Table 3.2. Further, there is an anti-reflecting coating on n-type material to minimize reflection of photons to increase absorptivity with electrical contact and glass to minimize long wave length radiation to escape from n-type material to ambient air from PV module. There is also back electrical contact below p-type material with polymer back sheet (tedlar). The glass and polymer back sheet plays an important role to avoid environmental contact with solar cell to its durability.

3.4.1 Photo-Voltaic (PV) Module

A solar cell with an effective area of 0.605 m^2 can produces 4.4 amp current (I_{sc}) with voltage of 500 mV (V_{oc}) under standard test condition (STC); 1000 W/m^2 and surrounding temperature of 25 °C under simulation condition as mentioned earlier, Table 3.2. The output of single solar cell will be $I_{sc} \times V_{oc} = 4.4 \times 0.5 = 2.2$ W. Now, in order to increase power, one can increase either current or voltage but in the case of solar cell current will depend on incoming solar energy and hence it is the oTo increase voltage, solar cells should be connected in series. Such a unit is known as a photo-voltaic module. If 36 solar cells are connected in series it will give 18 V,

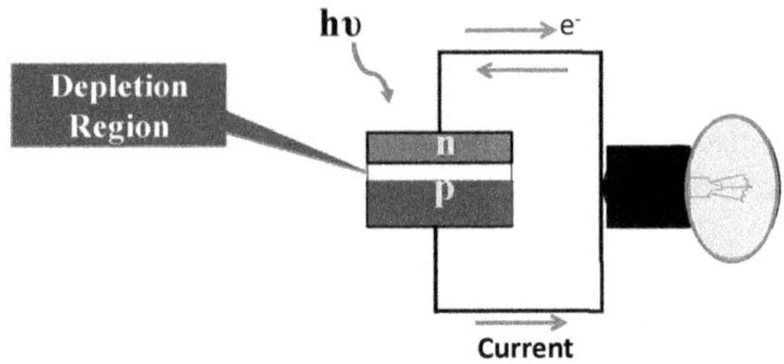

a

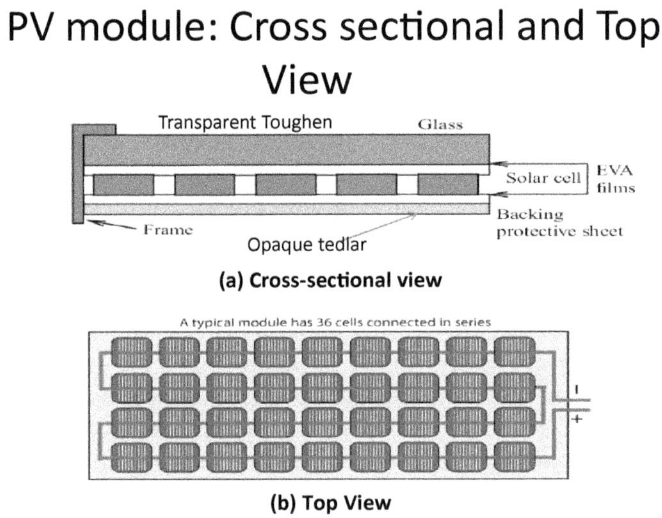

b

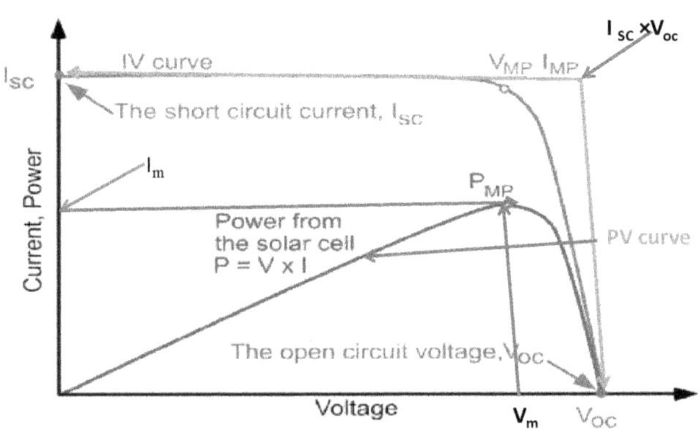

c

Fig. 3.4 **a** Flow of electron and current in solar cell outer circuit (working principle). **b** Pictorial view of cross-section and top view of PV. **c** I–V and P–V curve of a PV module

Table 3.2 Packing factor and specification of photo-voltaic (PV) module under standard test condition (STC)

S. No.	Variable parameters	Numerical value
1	Short circuit current, I_{sc}	4.4 amp
2	Open circuit voltage, V_{oc}	18 V
3	Solar intensity, I_0	1000 W/m^2
4	Area of one solar cell	0.605 m^2
5	Packing factor of PV module (PF)	0.89
6	Electrical efficiency, η_0	12%
y	Number of solar cell	36

with the same current of 4.4 amps under STC as given in Table 3.2. So, its voltage in outdoor (open) conditions will be below 18 V and hence it becomes easy to charge a battery of 12 V. The cross-section view and top view of PV module has been shown in Fig. 3.4b.

In this case, a packing factor (PF) is defined as follows:

$$PF = \frac{\text{Number of solar cells} \times \text{Area of one sole cell}}{\text{Area of } PV \text{ module}} \tag{3.2a}$$

The packing factor is one for rectangular shape of solar cell and less than one if there is spacing between two solar cells.

For a PV module, I–V and P–V curve has been shown in Fig. 3.4c. The short circuit current (I_{sc}) which is obtained by connecting two terminal of PV module without any load and open circuit voltage (V_{oc}) which is measured by opening both terminal without load by tong meter have also been shown in same figure.

Referring to Fig. 3.4c, a fill factor (FF) of a PV module can be defined by

$$FF = \frac{V_{MP} \times I_{MP}}{I_{sc} \times V_{oc}} \tag{3.2b}$$

The V_{MP} and I_{MP} are voltage and current at maximum power, Fig. 3.4c. The numerical value of FF is always less than one. The FF factor is defined as sharpness of I–V curve. Its value is provided by manufacturer of PV module.

The power from such combination without load will be calculated as

$$P_{MP} = FF \times I_{sc} \times V_{oc} \left(W_p\right) \tag{3.2c}$$

The above power is without load under standard test conditions. In actual open load conditions, the current varies with solar radiation.

There will be another maximum power with load which can be written as

$$P_m = I_m \times V_m \tag{3.2d}$$

The electrical efficiency of PV module depends on temperature of solar cell and its expression is given by

$$\eta_m = \eta_{m0}\left[1 - \gamma(T_c - T_{STC})\right] \tag{3.2e}$$

The 'γ' is power temperature coefficienct and it is equal to 0.0045 °C^{-1} and η_{m0} is electrical efficiency under STC.

The optimum inclination of PV system should be

Inclination of solar thermal system,

$$(\beta) = latitude(\varphi) \pm 15°, \tag{3.2f}$$

The plus (+) refers to winter conditions and minus (−) refers to summer conditions but for any PV system, it should be equal to latitude (φ).

Example 3.1 Calculate the number of solar cells for parameters given in Table 3.2 and corresponding power at normal, open conditions.

Solution: Given: open circuit voltage V_{oc} = 500 mV = 0.5 V and short circuit current I_{sc} = 4.4 amp at STC.

But in open conditions, the temperature of solar cells is higher than 25 °C and hence there is drop in voltage by 0.08 V.

At normal operating temperature, the voltage available across the terminals of each crystalline solar cell = 0.5–0,08 = 0.42 V.

Now, it is standard to make a solar module which can give 15 V open-circuit voltages to charge the battery of 12 V at any condition.

Hence, the required number of solar cells to construct such a solar module = $\frac{15}{0.42} = 35.71 \approx 36$,

Therefore, 36 crystalline (c–Si) solar cells are required to build a standard **solar PV module** of 15 V with maximum of 66W.

For a crystalline (c–Si) solar cell, the power = 15 × 4.4 = 66 W which is less than 75 W under STC

So, one can concludes that number of solar cell depends upon required voltage

If one requires more power then PV modules are connected in series to further increase the voltage as shown in Fig. 3.4c. This combination is referred as **string** of PV module. If string of PV module is connected in parallel to increase the current then it is referred as **array** of PV module.

To charge of combination of battery with given voltage, the actual voltage of PV module array should be at least higher by 15–20%.

Example 3.2 Calculate the number of PV module with specification of 18 V under STC to charge the lithium battery of 48 V.

Solution:

In normal open condition, PV module will give 15 V (Example 3.1).

To increase the voltage, PV module will be connected in series as shown in Fig. 3.5. If four PV modules are connected in series then it will give 60 V which will be sufficient to charge the 48 V lithium battery. If by any fault three PV modules are used then it will give 45 V which is not sufficient to charge the battery.

Example 3.3 Calculate electrical efficiency of PV module at 35 and 50 °C for data of Table 3.2 under STC = 25 °C.

Solution:

From Eq. 3.2e, we can get

$$\eta_m = \eta_{m0}\left[1 - \gamma(T_c - T_{STC})\right] = 0.12[1 - 0.0045(35 - 25)] = 0.1146 = 11.46\ \%$$

and,

$$\eta_m = 0.12[1 - 0.0045(50 - 25)] = 0.1065 = 10.65\ \%$$

One can observe that electrical efficiency decreases with an increase of solar cell temperature as expected.

The c-Si photo-voltaic (c–Si PV) module has captured about 80–85% global market share over the other PV modules (Thin film, CdTe, HIT, CIGS, a–Si), hence most people prefer to use silicon-based PV modules for various applications including domestic, water pumping, agriculture and street lights. The PV modules can be used in two modes:

(a) **Off-grid**: In this case, one needs a charge controller, inverter and battery bank which can be used for storage of electrical power for larger capacity of PV system $\geq 1kW_p$ due to storage of electrical power for safety. For a smaller capacity, one can escape the charge controller e.g. e-rickshaw and lighting. In this case, one needs to switch the charging unit off and on.

(b) **Online (net metering) system**: In the case of online (net metering), one does not require battery but electrical power in grid should be available during day time so net metering system should work. Online use of PV system is more economical then off-grid system. Roof top PV system is generally based on net metering system in all big cities where grid power is always available.

For higher voltage and power PV modules are connected in series as shown in Fig. 3.5. This is refer as PV panel system.

As solar radiation varies throughout the year, we can consider annual average solar radiation as $\overline{I}_{annual}(kWh/m^2)$. For a N–PV module with electrical efficiency of η_m (Eq. 3.2d) and effective area of A_m, an annual electrical energy from PV system will be determined by

$$E_{annual} = \overline{I}_{annual} \times N \times A_m \times \eta_m\ \text{kWh} \tag{3.3a}$$

If number of PV modules N and the wattage W_p are known, then annual electrical energy obtained from such PV system in kWh under STC can be evaluated as

$$E_{annual,STC} = \frac{N \times \text{Wp}}{1000} \times 365\ \text{kWh} \tag{3.3b}$$

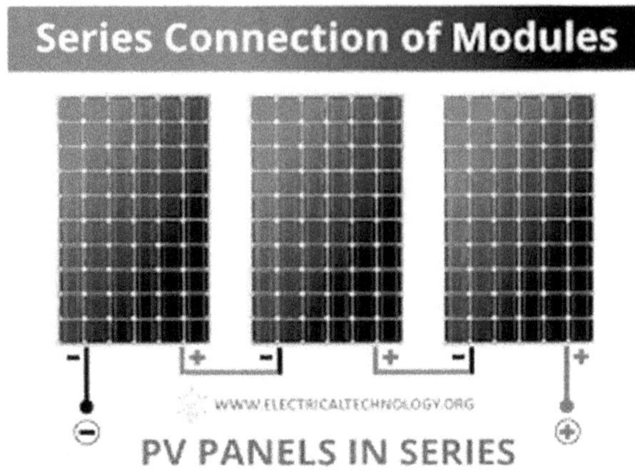

Fig. 3.5 Series connection of PV module (string)

For open conditions, it should be multiplied by 0.55, the above equation reduces to

$$E_{annual,open} = \frac{N \times \text{Wp}}{1000} \times 365 \times 0.55\,\text{kWh} \tag{3.3c}$$

The reason for multiplying by 0.55 is that in open/real condition an average solar radiation is about 550 W/m^2 as explained in Chap. 2.

Example 3.4 Evaluate the expected annual electrical energy production for the average annual solar radiation ($\overline{I}_{annual}$) of 1487 kWh/m^2 and PV module area of 120 × 1.67 m^2. with electrical efficiency of $\eta_m = 0.12$.

Solution:

From Eq. 3.3a, we have annual electrical energy

$$E_{annual} = \overline{I}_{annual} \times N \times \text{Am} \times \eta_m = 1487 \times 120 \times 1.67 \times 0.12 = 35,759.38\,kWh$$

Example 3.5 Calculate the expected annual energy production for 120 solar PV modules (N) with each of 200 W$_\text{p}$ under STC

Solution:

Fron Eq. 3.3c, we have annual electrical energy

$$E_{annual,open} = \frac{120 \times 200}{1000} \times 365 \times 0.55 = 4818\text{ kWh}$$

3.5 Wind Energy

Wind energy conversion systems (WECS) are made to convert moving wind (Sect. 2.6) with mass flow rate of $\dot{m}$ and density ρ carrying kinetic energy per second of $[\frac{1}{2}\dot{m}(u_0)^2 = \frac{1}{2}\rho A_1(u_0)^3]$ into mechanical power by wind turbine having cross-sectional area of A_1. Further, the mechanical power generated by the turbine is converted into electrical power by the generator. In fact, the generator is already integrated with the turbine. There are two types of wind turbine: horizontal axis and vertical axis. The most economical and popular turbine is horizontal axis type, and we will discuss the horizontal turbine which is most popular throughout world.

With reference to the wind cycle discussed in Sect. 2.6, it is clear that wind speed can vary with altitude. This variation has been shown in Fig. 3.6a. The optimum wind speed for extracting power from wind can be determined by considering average solar radiation of 500 W/m^2 equal to kinetic energy of incoming wind velocity as

$$\frac{1}{2}\rho(u_0)^3 = 500$$

or,

$$(u_0)^3 = \frac{2 \times 500}{\rho} = \frac{1000}{1.16} \quad \text{where} \quad \rho = 1.16\frac{\text{kg}}{\text{m}^3} \text{ at } 300\text{ K}$$

or,

$$u_0 = \left(\frac{1000}{1.16}\right)^{\frac{1}{3}} = 9.51 \text{ m/s}$$

This is incoming average minimum velocity of wind ($u_0 = 9.51\text{m/s}$) to be used for installation of wind energy conversion systems (WECS). This is very close to the maximum power point tracking (MPPT) system which occurs at 11 m/s as shown in Fig. 3.6a. The maximum power can be expected from WECS with wind velocity at 14 m/s. Beyond 14 m/s wind velocity, condition of turbulence is created hence power generation is badly affected.

Based on conservation of momentum theory, an expression for mechanical power extraction with incoming wind velocity of u_o for horizontal axis wind turbine (Fig. 3.6b) can be expressed as follows:

$$P_T = C_p\frac{1}{2}\rho A_1(u_0)^3 = C_p\frac{1}{2}\rho(\pi R^2)(u_0)^3 \tag{3.3}$$

where u_0 is incoming wind speed and $C_p = 0.59$ (maximum value) is power coefficient of horizontal axis wind turbine. If $C_p = 0$ then there is no power extraction from the wind, and its value varies between zero and 0.59 ($0 < C_p < 0.59$).

An expression for maximum electrical power generation from wind energy conversion system (WECS) can be written as

$$P_E = \eta_G \times C_p\frac{1}{2}\rho A_1(u_0)^3 = \eta_G \times P_T \tag{3.4}$$

where $\eta_G < 1$ and but ≥ 0.9 is generator efficiency.

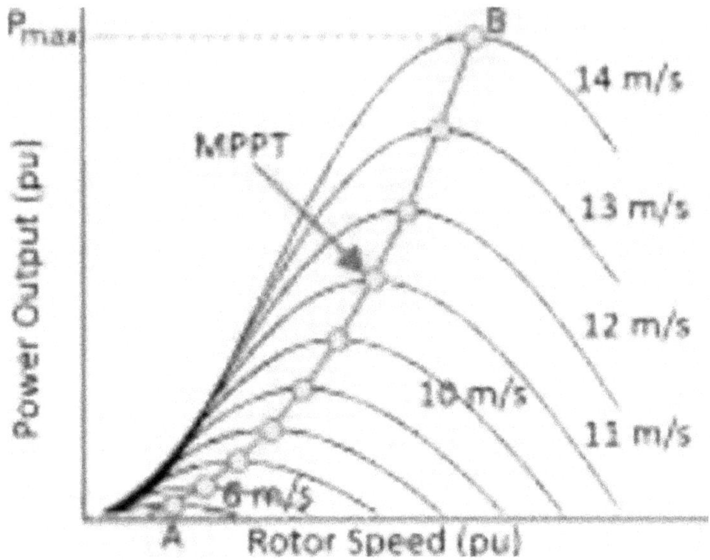

a

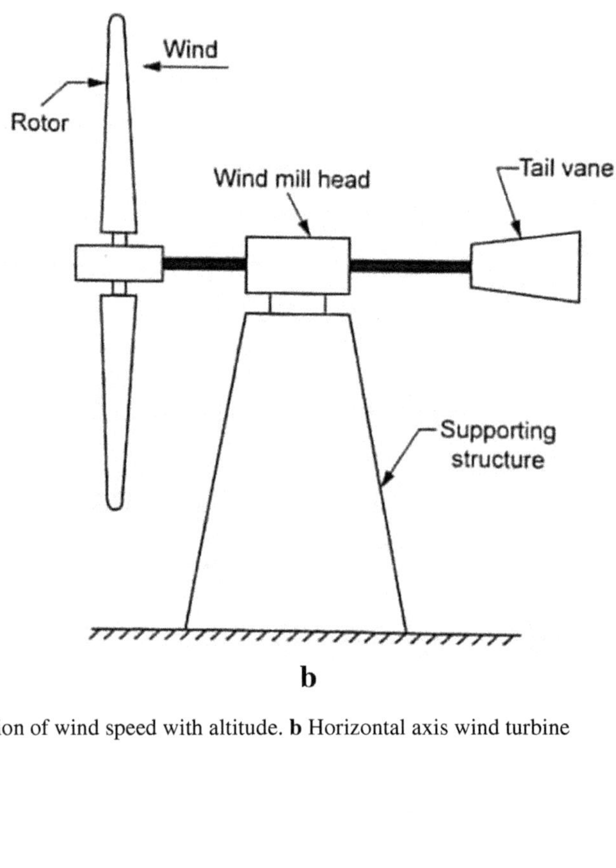

b

Fig. 3.6 **a** Variation of wind speed with altitude. **b** Horizontal axis wind turbine

3.5.1 *Advantages and Disadvantages of Wind Energy Conversion Systems (WECS)*

The followings are the advantages and disadvantages of wind turbine conversion systems (WECS):

Advantages are as follows:

(i) It is renewable and available free of cost in hilly and coastal regions.
(ii) It is helpful in supplying electric power to remote areas.
(iii) Wind power generation is cost effective and reliable.
(iv) It is pollution free, reliable and is an economical energy source.

Disadvantages are as follows:

(i) Wind energy has low energy density and is better in some geographical locations than others.
(ii) Requires energy storage batteries which indirectly and substantially contribute to environmental pollution.
(iii) Turbine rotors are not very efficient as they extract only 10 to 40% of the available wind energy due to the turbine power coefficient of 0.59.
(iv) Wind energy conversion system is capital intensive.

Example 3.6 Determine the mechanical power from the wind having wind speed of 20 m/s and blade length (radius of turbine, R) = 50 m for power coefficient (C_p) is equal to 0.59.

Solution:

Given parameters: Wind speed $u_0 = 20$ m/s, Blade length (R) = 50 m, Air density $\rho = 1.23$ kg/m^3.

The swept area $(A_1) = \pi R^2 = 3.14 \times 2500 = 7850$ m^2.

From Eq. 3.3 one has expression for mechanical power from wind energy as

$$P_T = C_p \frac{1}{2} \rho \left(\pi R^2\right) (u_0)^3$$

After substituting appropriate value, one gets

$$P_T = 0.59 \times \frac{1}{2} \times 1.23 \times 7850(20)^3 = 22,786,980\,\text{W} = 22,786.98\,\text{kW}$$

Example 3.7 Calculate mechanical power without wind turbine for Example 3.5

Solution:

In this case $C_p = 1$ and then

$$P_T = \frac{1}{2} \times 1.23 \times 7850(20)^3 = 38,622\ \text{kW}$$

So, the mechanical power obtained without wind turbine is

$$P_T = 38,622 \text{ W}$$

Example 3.8 Determine electrical power generation from WECS for Example 3.6 for electrical efficiency of generator = 0.9

Solution:

An expression for maximum electrical power generation from WECS from Eq. 3.4 is can be written as

$$P_E = \eta_G \times P_T = 0.9 \times 22,786.98 \text{ kW} = 20,508.282 \text{ kWh}$$

3.6 Hydroelectric Power

Hydroelectric power is renewable due to the hydro cycle (Sect. 2.2). The hydro cycle causes rain over hilly regions, and the rain water is stored at sufficient height (altitude) in natural dams, or reservoirs (see Fig. 3.7a as a potential energy source for hydro power. The construction costs of dams are significant. It is a form of energy obtained from water fall (intake) from a sufficient h, eight containing potential energy. In this case, potential energy (PE) of rain water stored at altitude of h is first converted in to kinetic energy.

The expression for potential energy is given by

Potential energy contained at height h = m × g × h (Joule)

The input power will be expressed as

Input power = $P_i = \frac{\mathrm{m \times g \times h}}{t} = \frac{\rho_w V g h}{t} = \rho Q g h$, W

$$P_i(kW) = \frac{\rho Q g h}{1000} \tag{3.5a}$$

where Q is volume flow rate, m^3/s, g = 9.8 m^2/s (acceleration due to gravity), h is the height of reservoir of water.

The expression for kinetic energy (KE) is given by

$$\begin{aligned} \text{Input kinetic energy (KE)} &= \frac{1}{2}\dot{m}(u_0)^2 = \frac{1}{2}\rho_w a u_0 (u_0)^2 \\ &= \frac{1}{2}\rho_w a (u_0)^3 = \frac{1}{2}\rho_w Q (u_0)^2 \end{aligned} \tag{3.5b}$$

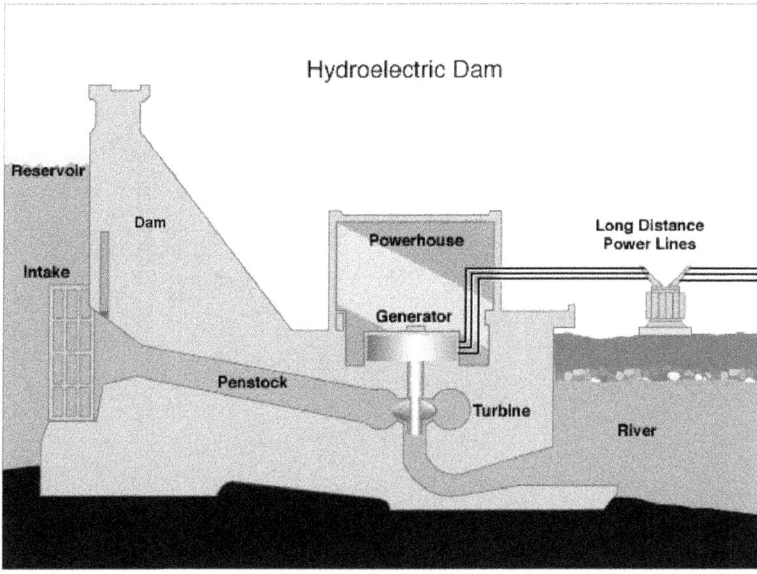

a

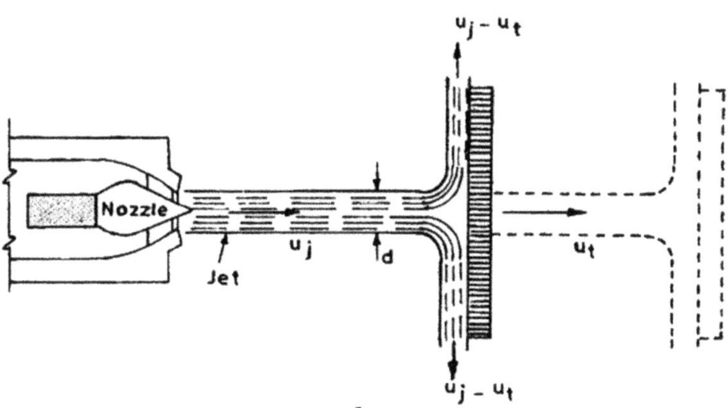

b

Fig. 3.7 **a** Hydro-power electrical generation system. **b** The water from nozzle is striking normal to vertical plate [6]. **c** View of water jet striking symmetrically a curved vane at its centre direction of Pelton wheel turbine (Fig. 3.7d) unlike Fig. 3.7b [6]. **d** Pelton wheel water turbine

where $\rho_w \approx 1\,000$ kg/m^3, Table 2.2 is density of water, **a** is cross section area of water fall, $\boldsymbol{u}_0$ is water velocity, m/s and **Q** is water volume flow rate, m^3/s.

By reducing cross-sectional area of water fall from 'a' to 'a_j' as per Bernoulli theorem$\left(au_0 = a_j u_j\right)$, the velocity of water fall, u_0, can be increased as water jet velocity$\left(u_j\right)$. Then the water jet strikes the blades of the Pelton wheel turbine moving with velocity of u_t to convert the kinetic energy into mechanical energy as shown in

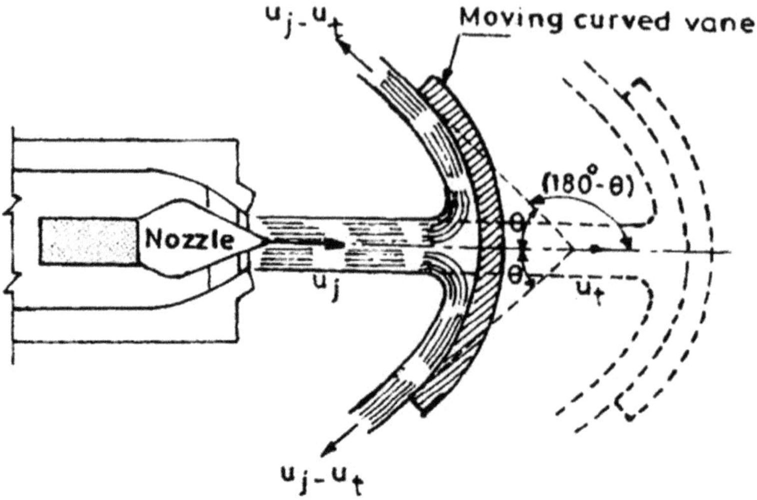

c

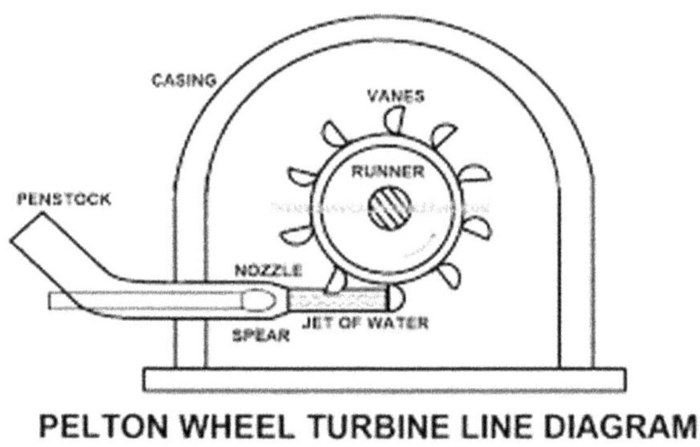

d

Fig. 3.7 (continued)

Fig. 3.7b. The turbine which converts kinetic energy of falling water into mechanical energy has been classified as a reaction and impulse turbine. Nowadays, the preferred turbine is the impulse turbine. There are many types of impulse turbine. The Pelton wheel turbine is the most efficient and easiest to install. The mechanical energy is transformed into electrical energy by the generator. People have used electrical power produced in this way for millennia. Over two thousand years ago, people in Greece used mechanical power obtained from flowing water to turn the wheels of mills to ground wheat into flour.

The input power associated with the Pelton wheel turbine [6] with moving blades with velocity 'u_t' is given by

$$\text{Input power } (P_i) = \frac{1}{2}\dot{m}(u_j)^2 = \frac{1}{2}\rho_w a_j (u_j)^3 = \frac{1}{2}\rho_w Q_j (u_j)^2 \tag{3.5c}$$

Referring to Fig. 3.7b, the mechanical power output from vertical plate moving with velocity, u_t by using the rate of change of momentum (force $\times velocity$) can be written as

$$\text{Out put mechanical power } P(u_t) = F \times u_t = m \times \frac{dv}{dt} \times u_t = \rho_w Q_j (u_j - u_t) u_t \tag{3.6}$$

The efficiency of hydropower (η) can be defined by using the Eqs. 3.5a, 3.5b, 3.5c and 3.6 as follows:

$$\eta = \frac{Out\ power}{Input\ power} = \frac{\rho_w Q_j (u_j - u_t) u_t}{\frac{1}{2}\rho_w Q_j (u_j)^2} = \frac{2(u_j - u_t) u_t}{(u_j)^2} \tag{3.7}$$

In order to optimize efficiency of hydropower (η), one has to differentiate Eq. 3.7 with respect to u_t for constant u_j and put equal to zero as

$$\frac{d\eta}{du_t} = \frac{2(u_j - 2u_t)}{(u_j)^2} = 0 \approx u_t = \frac{u_j}{2}\ or\ u_j = 2u_t \tag{3.7}$$

Substitute in the Eq. 3.6 then one has

$$\text{Out put mechanical power } P(u_t) = \rho_w Q_j (u_j - u_t) u_t = \frac{1}{4}\rho_w Q_j u_j^2 \tag{3.7b}$$

Substitute the above expression for u_t in Eq. 3.7, one gets
$\eta = \frac{2(u_j - u_t)u_t}{(u_j)^2} = \frac{1}{2} = 0.5$. For vertical plate attached to turbine, Fig. 3.7b [5].

So one can see that wind power conversion system (WECS) is more efficient (0.59) (Sect. 3.5) in comparison with hydropower with vertical plates.

Now, in the Pelton wheel turbine, there is a symmetrical curved plate as shown in Fig. 3.7c, d. An expression for maximum mechanical efficiency can be derived as

$$\eta_m = \frac{1}{2}(1 + cos\theta) \tag{3.8}$$

where θ is the angle shown in Fig. 3.7c, The derivation of the above equation has been given by Tiwari and Mishra [5].

Equation 3.7b will be modified for Pelton wheel turbine (Fig. 3.7c) as

$$\text{Output mechanical power } P(u_t) = \frac{1}{4}\rho_w Q_j u_j^2 (1 + \cos\theta) \tag{3.9a}$$

For $\theta = 180°$, $\eta_m = 1$ i.e. 100% mechanical efficiency of Pelton wheel turbine. But it noted that mechanical efficiency of best Pelton wheel turbine is 85% due to the following reasons:

(a) Nozzle losses.
(b) Mechanical friction losses.
(c) Friction losses inside curved bucket.
(d) Non-alignment of jet.

Due to this reason $u_t = 0.45\ \mathrm{u}_t$
Hence, an overall efficiency of hydropower plant can be defined as

$$\eta_o = \eta_m \times \eta_e \tag{3.9b}$$

where η_e is electrical efficiency of generator

Example 3.9 Derive an expression for mechanical power of Pelton wheel turbine for $\theta = 0$ and $180°$

Solution:
From Eq. 3.9a we have

$$\text{Out put mechanical power } \mathrm{P(u_t)} = \frac{1}{4}\rho_w Q_j u_j^2 (1 + \cos 180) = \frac{1}{4}\rho_w Q_j u_j^2 \times 0 = 0$$

For $\theta = 0$,

$$\mathrm{P(u_t)} = \frac{1}{4}\rho_w Q_j u_j^2 (1 + \cos 0) = \frac{1}{2}\rho_w Q_j u_j^2$$

For $\theta = 180°$

$$\mathrm{P(u_t)} = \frac{1}{4}\rho_w \theta_j u_j^2 (i - 1) = 0$$

Example 3.10 Find out an overall efficiency of hydropower plant.

Solution:
From Eq. 3.9b, one has

$$\eta_o = \eta_m \times \eta_e = 0.85 \times 0.95 = 0.8075$$

Example 3.11 Determine electrical and mechanical power generated from Pelton wheel turbine hydropower for the following given parameters:

$\rho_w = 1000\,\mathrm{kg/m^3}$, $Q_j = 5\,\mathrm{m^3/s}$, $u_j = 40\,\mathrm{m/s}$ and $\theta = 45°$, 1 MW = 10^6 W

Solution:
From Eq. 3.9a we have

$$P(u_t) = \frac{1}{4}\rho_w Q_j u_j^2 (1 + \cos 45) = \frac{1}{4} \times 1000 \times 5 \times 40^2 (1 + 0.525) = 183\,\mathrm{MW}$$

3.7 Geothermal Energy

Geothermal energy is the thermal produced deep in the Earth's core [7]. It originates from the formation of the planet and from radioactive decay of materials. It is thermal energy/heat which is derived from within the sub-surface of the earth to the Earth's surface as hot water/air/water steam. Depending on its characteristics and form, geothermal energy can be used either for space heating and cooling purposes or to be harnessed to generate clean electricity. As briefly mentioned about Earth in Sect. 1.3, the outer surface of Earth is about 288 K and its inner core in isothermal stage at about 6000 k which is equal to black body temperature of Sun. Geothermal energy (GE) is one of the most beneficial sources of energy.

Most economical ways to use geothermal energy as follows:

(i) Underground structure of building for thermal heating/cooling.
(ii) Earth air heat exchanger (EAHE) for space heating/cooling of building/ greenhouse in cold climatic condition. It is also useful in heating water for irrigation.
(iii) It is used in many industries such as food dehydration, milk pasteurizing and gold mining.
(iv) Hot water through springs, if available is directly sent to the buildings via pipelines in house hold for domestic use. It is used in Turkey for various domestic applications.

3.7.1 Advantages and Disadvantages of Geothermal Energy (GE)

(A) **Advantages**

(a) **Renewable Resource**: It is clean, renewable friendly with environment and can be produced 24 h a day.
(b) **Green energy**: It does not use any sort of combustion like underground fossil fuel (coal, petroleum, natural gas). GE emits only a small quantity of greenhouse gases.
(c) **Widely available**: The average temperature gradient near the Earth's surface is about 30 °C/km (0.03 °C/m). The internal heat of the earth is present globally. The extraction of thermal energy should be less than its availability.

(B) **Disadvantages**

(a) **Geothermal energy has high initial capital costs**: The digging wells to reach the geothermal reservoirs is expensive.
(b) **Suited to a particular region**: It is hard to get the right type of hot rocks near cities. The rocks should also be found at a lower depth to make drilling easier.

(c) **Sustainability issues:** GE can be an issue if hot water is removed faster than it can be replenished.
(d) **Transportation and transmission**: Unlike fossil fuels, geothermal energy cannot be directly transported from one place to another place.

3.8 Human Energy

Human energy is the rate of work per unit time produced from the human body. It is one of renewable energy source which is also friendly with environment and climate. It is also referred as human power. Human power comes primarily from muscles. Normal human metabolism processes in body produces heat at a basal metabolic rate of around 80 W to do work like warming closed shelters or other humans.

Human power can be used for the following [8]:

(a) **Transport**: Several forms of transport powered by human effort are the bicycle, Fig. 3.8a, wheelchair, walking, skateboard, wheelbarrow, rowing, skis, rickshaw, swimming, pedal-boat and many other forms of mobility. During a bicycle race, 400 W of mechanical power an hour is produced by a cyclist. Bicycle riding, Fig. 3.8a, has five advantages: [7 friendly to the air and climate, no pollution, no parking problems, robust and renewable.
(b) **Exercise/yoga**: During exercise, an average between 50 and 150 W per hour is produced by heat.
(c) **Manual healthy laborer**: Their output of around 75 W of power has been produced by working of eight hours in all sectors including agriculture.
(d) **Pedal-powered boat**: Humans have used a 26-foot-long pedal-powered boat to cross the oceans, Fig. 3.8b.
(e) **Wristwatches**: In this case, muscle power is used to keep their mainsprings wound up.
(f) **Tread wheels**: Also known as treadmills. It is a machine powered by humans for mechanical energy.
(g) **Electrical power**: Human-powered equipment primarily consists of electrical appliances powered by human muscle as an alternative to conventional sources of electricity such as disposable primary batteries and the electrical grid.
(h) **Communication**: It is a common application for the relatively small amount of electric power that can be generated by a human muscle turning a generator.

a

b

Fig. 3.8 **a** Peddler cycle riding. **b** Peddler boat roding

3.8.1 Busting Immune System

In order to increase human muscle power, one has to the immune system [8]. This can be achieved by the followings:

(i) **Sun bathing:**: Following Sect. 3.2 and Fig. 3.1, one should sun bathe bath early in the morning before 8 am to absorb vitamin D. Sun bathing not only helps the body to absorb vitamin D naturally, but also improves sleep, **strengthens bones and boosts the immune system.** Vitamin D deficiency is present in around one billion people globally. Solar energy from sun provides energy to all living organisms to sustain ecology, environment and climate. Some countries are very fortunate to have sufficient solar energy like India, whereas others have to create artificial solar energy, like Sweden.

(ii) **Warm water**: One should use warm (54–71 °C) fresh underground water for drinking /washing/ bathing if freely available. Drinking water, hot or cold, keeps your body healthy and hydrated. Consuming hot water maintains health and hydration and offers improvement of digestion, relieves congestion and promotes relaxation in comparison to cold water. Warm water can also be obtained by using various solar energy devices [9].

(iii) **Physical activities (Exercise)**: Exercise means the movement of body. It helps your muscles work by burning calories, enables weight loss, increases energy level, improves sleep and lengthens life. The most beautiful movement of the body can be seen in newly born baby in presence of solar energy in the morning. No one tells them about movement of the body, but they satisfy their need automatically. Simple physical activities include morning and evening walks, running, cycling, swimming, playing games and dancing. Due to being physically active, working people have fewer health problems than others.

(iv) **Parents and senior citizen**: Parents (father and mother) give birth to their children and they take care of them. They play **important** roles in growing, walking, physical movement, eating, etc. in childhood and work hard to meet the requirements of finance and career development for the future along with happiness with good mental, physical, and social activities. So, one should be proud to have **parents**. It is also well known that peace of mind helps to control many health issues including happiness, low tension, normal blood pressure and longer life etc. To achieve this, one should take care of living parents if possible.

(v) **Inner thoughts**: The conscious experience of an individual human being should be regarded as a continuous positive inner voice and ideas running through the self -mind. It should be as maximum as possible. It give peace of mind for longevity of life.

(vi) **Medicial plants**: Consume regularly green/fresh leaf of holi tulsi (**Ocimum tenuiflorum/** Basil), neem (**Azadirachta indica**) neem (**Azadirachta indica**), guava, hibiscus (gudhal), mint, curry etc. on an empty stomach in the morning if possible. These have many medicinal properties which include strengthening the stomach and helping in respiratory diseases and kidney stones. Consumption of these controls many diseases in advance such as stomach upset, heart and blood vessel issues, eye disorders, intestinal worms, skin ulcers, fever, diabetes, and liver problems.

(vii) Encourage people to keep their surroundings clean and stay within the natural habitat.

3.8.2 Energy Released by Human

The human body functions from (a) thinking to lifting weights and (b) quiet activity, sleeping to head scratching [8]. These needs energy and small visible muscle actions (heart, lungs, and digestive tract). The metabolic rate is the rate at which the body uses food energy to sustain life. The total energy conversion rate of a individual human at rest is known as the basal metabolic rate (BMR) and it is divided among various systems in the body as given in Table 3.3. One can observe that the liver and spleen consume maximum oxygen (O_2) while the heart consumes minimum oxygen (O_2). The largest fraction of energy goes to the liver and spleen, and the minimum goes to the heart. Nearly 75% of the calories burned in a day go into the liver and spleen and skeletal muscle. A full 25% of all basal metabolic energy consumed by the body is used to maintain electrical potentials in all living cells (nerve impulses).

Example 3.12 Calculate how much thermal energy is transferred to the environment by the person of a 65 kg climbs a 15 m high through stairs as exhaust heat? Consider the following assumptions:

(a) A 20% $\left[\frac{1}{5}\right]$ mechanical efficiency in climbing stairs and
(b) The exhaust heat is $\left[\frac{4}{5}\right]$ of the total loss of internal energy.

Solution:

The change in gravitational potential energy $\left(\Delta PE_g\right)$ which is actually work done by a person (W) is given by

$$\Delta PE_g(W) = mg\Delta h = 65 \times 9.8 \times 15 = 9,555\text{J}$$

Table 3.3 Basal metabolic rates (BMR) for human [10]

S. No	Organ	Power consumed at rest (W)	Oxygen (O_2) consumption (mL/min)	BMR (%)
1	Liver and spleen	23	67	27
2	Skeletal muscle	15	45	18
3	Kidney	9	26	10
4	Heart	6	17	7
5	Brain	16	47	19
7	Other	16	48	19
8	**Totals**	85	250	100

The person converts chemical potential energy in their body to gravitational mechanical energy during any physical work.

The person did work in converting chemical potential energy in their body to mechanical energy, specifically gravitational potential energy. However, they are only 20% efficient.

As per first assumption, the 1/5 of the chemical potential energy can be used for doing useful work. Therefore, the total chemical energy in body of a person will be five times the change in gravitational potential energy. Hence,

The total chemical energy in body of a person $= 5 \times \Delta PE_g = 5 \times 9,555 = 47,775$ J .

The chemical potential energy used came out of the person's internal energy (ΔU) so

$$\Delta U = -47,775 \text{ J}.$$

As per the First Law of Thermodynamics, we know that to find the thermal energy exhausted by the person:

$$\Delta U = Q - W or Q = \Delta U + W = -47,775 + 9,555 = -38,220 \text{ J}$$

Here, one can see that the heat is negative. This means the person exhausts thermal energy out of the body and into to the environment during climbing the stairs.

3.8.3 Energy Factor for Different Food [9]

Table 3.4 provides the energy factors applied to various foods, namely eggs, meat products, milk products, fats—separated, fruits, grain products, legumes, nuts and vegetables. Based on Table 3.4, one can observe the following:

Table 3.4 At water specific factors for selected foods/diet [11]

Food/diet	Protein (kJ/g)	Fat (kJ/g)	Total carbohydrate (kJ/g)
(a) Eggs, meat products, milk products			
Eggs	18.2	37.7	15.4
Meat/fish	17.9	37.7	—
Milk/milk products	17.9	36.8	16.2
(b) Fats—separated			
Butter	17.9	36.8	16.2
Margarine, vegetable	17.9	37.0	16.2
Other vegetable fats and oils	–	37.0	–
(c) Fruits			
All, except lemons, limes	14.1	35.0	15.1
Fruit juice, except lemon, lime#	14.1	35.0	15.1
Lemon, limes	14.1	35.0	10.4
Lemon juice, lime juice#	14.1	35.0	11.3
(d) Grain products			
Barley, pearled	14.9	35.0	16.5
Cornmeal, whole ground	11.4	35.0	16.9
Macaroni, spaghetti	16.4	35.0	17.2
Oatmeal—rolled oats	14.5	35.0	17.2
Rice, brown	14.3	35.0	17.2
Rice, white or polished	16.0	35.0	17.4
Rye flour—whole grain	12.8	35.0	16.2
Rye flour—light	14.3	35.0	17.0
Sorghum—whole meal	3.8	35.0	16.9
Wheat—97–100% extraction	14.0	35.0	15.8
Wheat—70–74% extraction	17.0	35.0	17.2
Other cereals—refined	16.2	35.0	17.2
(e) Legumes, nuts			
Mature dry beans, peas, nuts			
Soybeans			
(f) Vegetables			
Potatoes, starchy roots	11.6	35.0	16.9
Other underground crops	11.6	35.0	16.1
Other vegetables	10.2	35.0	14.9

(i) For ordinary food/diet, energy conversion factors for protein in the Atwater specific factor system vary 10.2 kJ/g (vegetables) to 18.2 kJ/g (egg). The percentage variation is $\frac{18.2-10.2}{18.2} \times 100\% = 44\%$.
(ii) Fat varies from 35 kJ/g **[c to f)** to 37.7 kJ/g **[a to b]**, The percentage variation is $\frac{37.7-35}{37.7} \times 100\% = 7\%$.

and
(iii) For total carbohydrate from 11.3 kJ/g (lemon and lime juices) to 17.4 kJ/g (polished rice.). The percentage variation is $\frac{17.4-11.3}{17.4} \times 100\% = 35\%$.

These ranges for protein, fat and carbohydrate are 44, 7 and 35%. Merrill and Watt [9]. Here, it is important to note that an individual's dietary needs for survival and to produce energy inside the body by metabolic processes are as mentioned earlier.

3.9 Animal Energy

In animals, carbohydrate [$(CH_2O)_n$] acts as the main energy source. The carbohydrate is broken down to glucose [$C_6H_{12}O_6$ n = 6 in $(CH_2O)_n$]. The glucose enters the metabolic pathways in animal and releases ATP (adenosine triphosphate). ATP is energy-carrying molecule found in the cells of all living things on the planet Earth including human and animals which is the main energy source.

The carbohydrate [$(CH_2O)_n$] is formed in the plants containing maximum water molecules (H_2O) by interacting with photon (light) of solar energy by photosynthesis process (Sect. 3.3, Fig. 3.3). The solar energy coming from the sun is at the heart of the cycle of life of all living organism. It is referred to as the food web. So, the Sun is the ultimate source of energy for animals too. Its food energy must be harnessed through the plants before it can be used by animals. Animals get energy by processing complex foods.

Animals can also be classified in three categories as follows:

(a) **Herbivores (vegetarian)**: These are the animals at the bottom of the animal food chain and eat plants to survive. The available nutrition for animals from grass and other green plants are called herbivores. Any kind of food is stored energy and essential elements like carbon, nitrogen and hydrogen. The herbivores (vegetarian) animals have digestive systems that allow them to process all types of green plants (grasses), tree bark, aquatic vegetation, and shrubby growth. Examples include cows, buffalo, elk, sheep, giraffes, donkeys, mules, horses, rabbits, chipmunks, squirrels, mice and elephants. Herbivores require an abundant intake of energy to stay alive. Many herbivores (horses, donkeys, cows, buffalos and camels) eat grass all day. Herbivores can't digest cellulose for humans.
(b) **Carnivores (non-vegetarian)**: Carnivores (non-vegetarian) animals eat other animals that are the last step in the food chain. They are the farthest away from

the ultimate source of energy. Predators like lions, wolves, hawks etc. that kill/hunt/prey other animal and eat their meat are known as carnivores animal.

(c) **Omnivores**: An animal that has the ability to eat and survive by using both plant and animal matter as a food. Such animals are omnivores, including pigs, dogs, bears, coatis, hedgehogs, opossum, chimpanzees, squirrels, raccoons and chipmunks. Carnivores' existence is necessary in the ecosystem because they prevent overpopulation of other animals. Further, they are at the least efficient end of energy-use in the food chain.

3.9.1 Working Animals

Working animals can used for

(a) **Transportation**: Pack animals, such as donkeys/asses, camels, oxen, reindeers, elephants, llamas, sheep, goats, yaks, and dogs are used to carry freight, goods, or supplies. The ass/donkey is the oldest-known pack animal, used as early as 3500 BC. In terrain area where wheeled vehicles would encounter difficulty. For example, camels, Fig. 3.9a and horses are used in the desert and mountainous terrain respectively. In many places in the world, the pack animals are the only used for transporting a load.

(b) **Agriculture**: Draught animals (DAs) are generally allowed to be used for agriculture operations namely tillage, seedbed preparation, sowing, weeding, and harvesting, threshing and post-harvest operations etc. They also improve human welfare after taking care of them in early years after human civilization. Draught animals assist humans in agricultural tasks (see Fig. 3.9b). In agriculture, the field is ploughed using bulls, oxen, cows, etc. The average mechanical power delivered by a pair of bullocks is about 1 hp for farm operations.

(c) **Butter churn**: It is a device for converting cream into butter. This is frequently done through a mechanical process. A wooden pole inserted through the lid of the churn to turn a rotating device inside the churn. The system is also known as a small treadmill intended for small animals like dogs and goats.

(d) **Oil extraction**: The wooden machine is extracting oil (ghani oil) from mustard seed is an ancient Indian technique (the cold-pressed way) as shown in Fig. 3.9c (i). Mustard husk can be used as food for aquaculture, briquettes (biomass), Fig. 3.9c (ii) etc.

Animal power is renewable, sustainable, and friendly to the environment and climate.

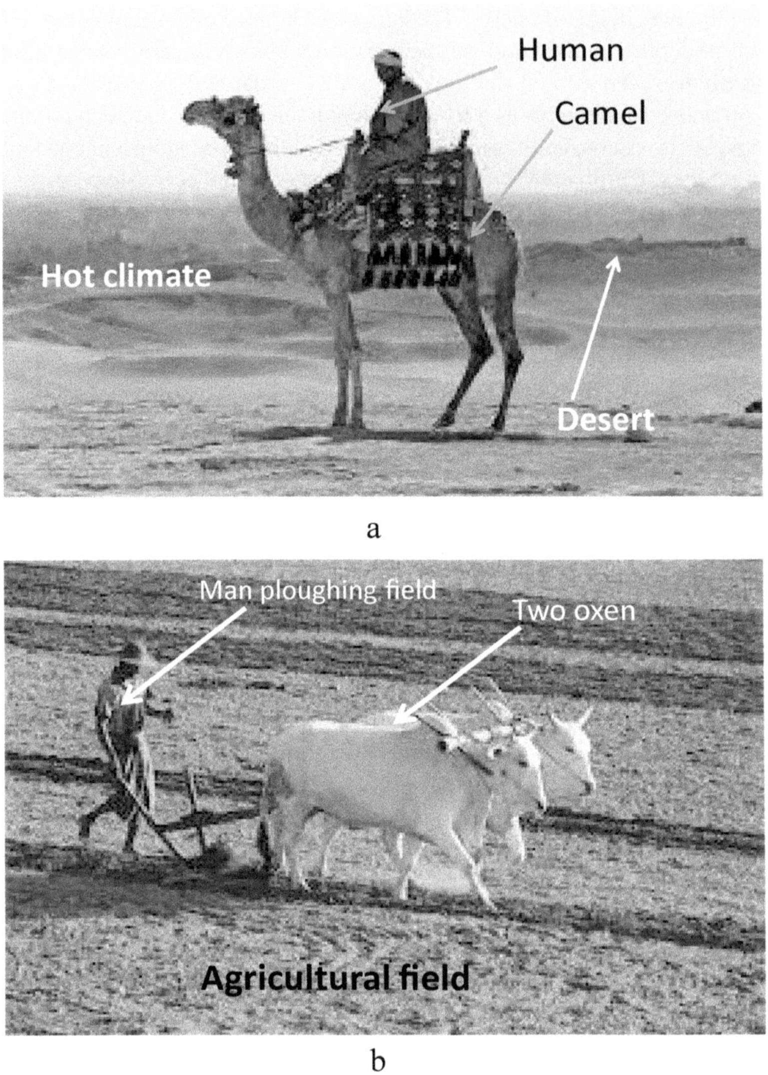

Fig. 3.9 **a** Camel (Paak) for transportation in sandy desert. **b** Ploughing agricultural field by oxen. **c** (i) Extraction of oil from oil seed by kohlu machine and (ii) mustard briquettes

Problems

Problem 3.1 Draw the plot of η_m with variation of solar cell temperature varying from 25 °C to 89 °C.

Hint: See Example 3.3

(i)

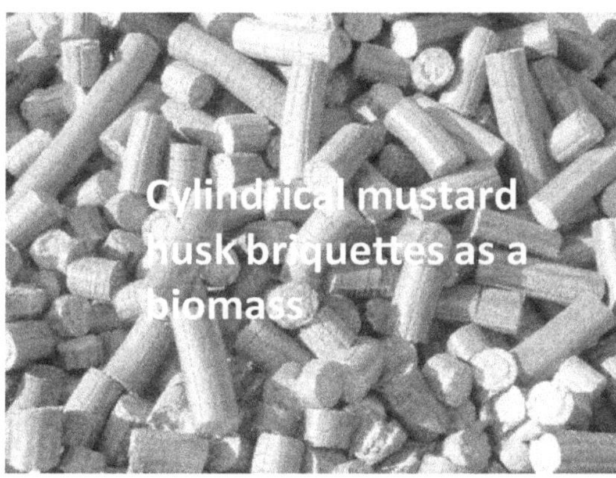

(ii)

c

Fig. 3.9 (continued)

Problem 3.2 Find the expected annual electrical energy production for the average annual solar radiation ($\overline{I}_{annual}$) of 1,200, 1,400 and 1,600 kWh/m^2 and PV module area of 120 × 1.67 m^2 with electrical efficiency of $\eta_m = 0.12$.

Hint: Use Example 3.4

Problem 3.3 Evaluate the expected annual energy production for 100, 200 and 300 solar PV modules (N) with each of 200 and 300 W_p under STC

Solution: Use Hint 3.5

Problem 3.4 Plot mechanical power of wind energy conversion system (WECS) with power coefficient from 0 to 0.59

Hint: See Example 3.6

Problem 3.6 Calculate wind speed for mechanical power of wind turbine as 22, 786.98kW.

Hint: See Example 3.6

Problem 3.7 Plot the curve between mechanical power $P(u_t)$ and angle θ for Example 3.9

Hint: Vary θ from zero to 180°.

Problem 3.8 Calculate volume flow rate (Q) for the flowing parameters of

$$\rho_w = 1000\,\text{kg}/\,\text{m}^3\; Q_j = 5\,\text{m}^3/\,\text{s},\, u_j = 40\,\text{m}/\,\text{s}$$
$$\text{and}\,\theta = 45^\circ,\; P(u_t) = 183\,\text{MW}$$

Hint: See Example 3.9

Problem 3.9 Determine electrical and mechanical power generated from ideal Pelton wheel turbine hydropower for Example 3.9

Hint: Consider $\theta = 0^\circ$

Problem 3.10 Determine how much thermal energy is transferred to the environment by the person of a 65 kg climbs a 15 m high through stairs as exhaust heat by considering a 20% $\left[\frac{1}{5}\right]$ mechanical efficiency during climbing stairs

Hint: First calculate the total chemical potential energy in body (see Example 3.10) which is the person's internal energy (ΔU) as 47, 775J and then multiply 4/5 with negative sign as explained in same example.

Objective Questions

3.1 The vitamin D from sun bath is good with

(a) Low shadow of human exposed to the Sun
(b) Large shadow of human exposed to the Sun
(c) Equal shadow of human exposed to the Sun
(d) None of these

Answer: (a)

3.2 Small shadow during sun bathing takes place at

(a) Morning
(b) Afternoon
(c) Noon
(d) Night

Answer: (c)

3.3 Optimum time for sunbathing at noon is

(a) One hour
(b) 3 h
(c) 15–30 min
(d) Two hours

Answer: (c)

3.4 The unit of vitamin D is

(a) mcg
(b) microgram
(c) IU
(d) None

Answer: (a-c)

3.5 Time for consuming Vitamin D in the morning

(a) More than evening period
(b) Equal to evening period
(c) More than noon period
(d) None of these

Answer: (c)

3.6 The plant during photosynthesis produces

(a) Hydrogen (H_2)
(b) Oxygen (O_2)
(c) Carbon dioxide (CO_2)
(d) All of these

Answer: (a) and (b)

3.7 The plant during photosynthesis consumes

(a) Hydrogen (H_2)
(b) Oxygen (O_2)
(c) Carbon dioxide (CO_2)
(d) All of these

Answer: (c)

3.8 The electrical efficiency of PV module

(a) Increases with temperature
(b) Independent of temperature
(c) Decreases with temperature
(d) None of them

Answer: (c)

3.9 The wind power from WECS is

(a) Inversely proportional to square of radius of turbine
(b) Proportional to square of wind velocity $(u_0)^2$
(c) Proportional to cube of wind velocity $(u_0)^3$
(d) All of these

Answer: (c)

3.10 The wind power from WECS is

(a) Inversely proportional to square of radius of turbine $(R)^2$
(b) Proportional to square of wind velocity $(u_0)^2$
(c) Proportional to cube of wind velocity $(u_0)^3$
(d) Proportional to square of radius of wind turbine $(R)^2$

Answer: (d)

3.11 The mechanical power of WECS is

(a) Proportional to electrical efficiency of generator
(b) Inversely proportional to generator electrical efficiency
(c) Approximately equal to electrical energy
(d) None of these

Answer: (a) and (c)

3.12 Mechanical power from WECS is

(a) Proportional to electrical power
(b) Inversely proportional to electrical efficiency of generator
(c) Equal to electrical power
(d) None of these

Answer: (a) and (b)

3.13 Electrical power from WECS is

(a) Proportional to mechanical power
(b) Proportional to electrical efficiency of generator
(c) Equal to mechanical power
(d) None of these

Answer (a) and (b)

3.14 Ideal mechanical power of Pelton wheel turbine is

(a) 50%
(b) 75%
(c) 100%
(d) 85%

Answer: (c)

3.15 An overall efficiency of hydropower plant is

(a) 50%
(b) 75%
(c) 100%
(d) 85%

Answer: (d)

3.16 Input mechanical power of a hydropower plant is

(a) Proportion to initial velocity (u_0)
(b) Proportional to square of initial velocity $(u_0)^2$
(c) Proportional to cube of initial velocity $(u_0)^3$
(d) None of these

Answer: (c)

3.17 Input mechanical power of hydropower depends on

(a) Volume flow rate (Q)
(b) Square of initial velocity, $(u_0)^2$
(c) Cross section area of jet
(d) All of these

Answer: (d)

3.18 Geothermal energy is extracted from

(a) Inside earth
(b) Inside sun
(c) The sky
(d) The sea

Answer: (a)

3.19 Geothermal electrical power generation

(a) Electric generator is used
(b) Mechanical turbine is used
(c) Steam generator is used
(d) All of these

Answer: (c)

3.20 Maximum protein is available in

(a) Eggs, meat products, milk products
(b) Fats—separated
(c) Vegetables
(d) All of these

Answer: (a) and (b)

3.21 Maximum fat is available in

(a) Eggs, meat products, milk products
(b) Fats—separated
(c) Vegetables
(d) All of these

Answer: (a) and (b)

References

1. Sun Bath Therapy: All You Need to Know About Sunbath. https://timesofindia.indiatimes.com
2. Sunlight and Vitamin D—PMC—NCBI. https://www.ncbi.nlm.nih.gov›articles›PMC3897598
3. CIE Technical Report. Action spectrum for the production of previtamin D3 in human skin. CIE 174:2006. ISBN 3901906509 (2006). 13. McKinlay, A.F., 5
4. B.L. Diffey, A reference action spectrum for ultraviolet induced erythema in human skin, in *Human Exposure to Ultraviolet Radiation: Risks and Regulations* (Elsevier, Amsterdam, 1987), pp. 83–87
5. National Institutes of Health (NIH) I Turning Discovery. https://www.nih.gov
6. G. N. Tiwari, R.K. Mishra, *Advanced Renewable Energy Sources* (Royal Society of Chemistry (RSC), (UK), 2011)
7. Five (5) Health and Environmental Benefits of Riding Your Bike. https://biofriendlyplanet.com
8. G.N. *Tiwari, Ecology, Environment and Human Being*, Published: 30 September 2020, by Juniper Publishers, Ecology & Conservation Science: Open Access, Volume 1; https://doi.org/10.19080/ecoa.2020.01.555565
9. G.N. Tiwari, *Arvind Tiwari and Shyam* (Springer, Handbook of Solar Energy, 2016)
10. Efficiency of the Human Body pressbooks.pub.·https://o
11. A.L. Merrill, B.K. Watt, *Energy value of foods: basis and derivation.* Agriculture Handbook No. 74. (ARS United States Department of Agriculture, Washington, DC, 1973)

Chapter 4
Energy Conservation

4.1 Introduction

All living organisms including wildlife, pet animals and humans etc. need basic energy provided by nature [1]. In fact, the Sun, which is a part of nature, provides energy directly or indirectly to all living things as discussed in Chapter-3. Most of this energy is absorbed by

(i) **Plants:** In the form of photosynthesis to produce food required by all living organism.
(ii) **Human:** In required as Vitamin D for human.

Energy is available in different forms. Only electrical energy (high grade) and thermal energy (biomass fuel; low grade) can be saved. Excessive use of electrical and thermal energy results in pollution of the environment (CO_2) and eradication of natural resources. The electrical energy provider is less efficient than thermal energy providers, hence electrical energy generation pollutes more in comparison to thermal energy. This imbalances nature. To conserve energy to sustain environment and nature, one uses as little as possible to meet requirements on a daily basis [2].

Nowadays, an electrical energy is available by using either renewable energy sources or non-renewable sources (fossil fuel: lignite, coal, petroleum and natural gas). It is well known that available electrical energy by renewable energy is friendlier to the environment and climate than those obtained from non-renewable sources such as fossil fuels which are limited sources. Table 4.1 suggests that lignite and hard coal emit much more CO_2 in environment in comparison with natural gas and petroleum at source of electrical power generation. This emission is further increased by transmission and distribution (T & D) losses per kWh. So, our first step in direction of energy conservation should be use renewable energy sources for both thermal as well as electrical energy for our day to day basic activities to meet our basic need. It should be understood by humans that nature has already provided for the basic needs of humans to live comfortably and be friendly to the environment.

G. N. Tiwari, *Energy, Ecology and Environment*,
https://doi.org/10.1007/978-981-99-3997-8_4

Table 4.1 Specific CO_2 emissions from fossil fuels in kg/kWh$_e$ [2]

S.no	Name of fuel	Mean electrical efficiency (%)	Kg CO_2/kWh$_{PE}$
1	Lignite	38	1.093
2	Hard coal	39	1.001
3	Natural gas	56.1	0.443
4	Petroleum (Diesel)	≈ 40	0.8–0.93

These can be achieved in many ways, such as the following:

(i) **Hot water**: Use of solar derived hot water for domestic tasks (cooking, washing clothes and floors, bathing, drinking etc.) and personal use. It can be directly as well as indirectly used.
(ii) **Electrical power**: Use of electrical energy obtained by PV system/biomass (rice husk/bagasse)/wind/hybrid in the form of DC power if possible by using DC appliances.
(iii) **Turning off the lights**: It should be done by using sensors related to human activities in required places.
(iv) **Removing plugged-in appliances**: It can be done manually if electrical appliances are not in use due to reverse flow of current which is responsible for energy losses.
(v) **Walking**: For small distances, one should prefer to either walk or cycle instead of using vehicles that emit CO_2 into environment and cost money. In doing so, there are many benefits to health which reduce the need for medicine, and it is better for the environment, and saves time and money.
(vi) **Sunbathe**: Early in the morning, it is encouraged to sunbathe for consumption of vitamin D directly from the sun to avoid the need to use vitamin D supplements.
(vii) **Use of energy efficient electrical equipment**: We should buy higher star-rated electrical appliances to consume less electrical energy if one can afford. For example, Compact Fluorescent Lamps (CFL) and Light Emitting Diode(LEDs) bulbs instead of regular incandescent bulbs.
(viii) **Construction of passive house**: We should use maximum exterior double glass walls/windows to trap sunlight in our homes, schools, and workplaces during the daytime for thermal heating in cold climatic condition. This will reduce our electricity bills. For cooling, we should use single glass windows as minimally for east/west/north side only for day lighting and cross ventilation in hot climatic conditions.
(ix) **Services electrical gadgets annually**: Annual services of fridges, air conditioners, coolers, domestic appliances, electrical machines, generators etc. results in better efficiency.
(x) **Use of organic food/medicinal plants**: It helps immunity of humans to fight against diseases due to changes in weather conditions.

(xi) **Use of community facility**: In this case, electrical consumption for any general activities is reduced up to certain level.
(xii) **Use of any device/units/system with low energy density materials**: It helps the reduction of input energy to prepare any tools for further use.
(xiii) **Population growth**: Limiting population growth, particularly in developing and under developing countries can help reduce energy consumption.
(xiv) **Green energy**: Renewable energy including human energy. Mankind should use direct green energy to meet daily requirements, particularly in the agricultural sector as mentioned earlier.
(xv) **Green engineering practices**: It is possible to minimize losses due to friction of the components of various machines by oiling to improve the life span of the components.

The above points to be taken care by all by using thermal energy which is available freely from the Sun and biomass, and electricity from hydropower etc.

In the recent past, particularly after World War Two, electrical power has been generated by using fossil fuels, which are limited and not friendly to the environment and climate. The carbon dioxide (CO_2) emissions by different fossil fuels for electrical power generation at source is given in Table 4.1. It is to be noted that CO_2 emission per kWh energy generation is maximum for lignite, hard coal and petroleum ($\approx$ 1kg/kWh) in comparison with natural gas ($\approx$ 1kg/kWh). Hence natural gas is preferred due to its lower emission of CO_2. This emission is further increases if transmission and distribution of electrical power is considered, which will be discussed later.

In the next section, energy (kWh) per capita, which includes both renewable (solar and biomass fuel) and non-renewable (fossil fuels), will be discussed.

4.2 Energy Consumption (kWh) Per Capita

It is important to mention energy consumption (kWh) per capita for some developed, developing and under developing countries. These data are given in Table 4.2 with reference to the years 2014 and 2022 [3] respectively. The average global energy consumption (kWh) per capita is 19,836 kWh [4]. The Total energy consumption per capita for country in the year 2019 [5] has been shown in Fig. (4.1).

The condition to become developed, under developing and under (least) developing countries is as follows.

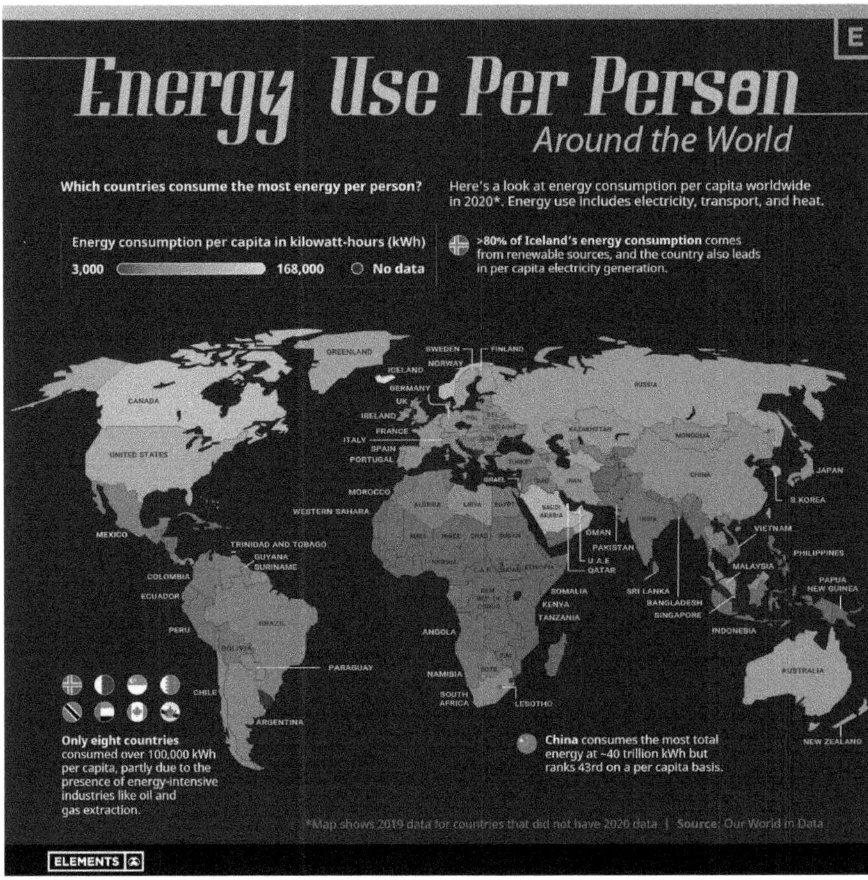

Fig. 4.1 Total energy consumption per capita for each countries in the year 2019 [5]

4.2.1 Developed Countries

(a) **Human development index (HDI)**: It accounts for more than net income or productivity. Further, how income is turned "into education and health opportunities and therefore into higher levels of human development."
(b) **High-income economies**: It depends on GDP growth per capita.
(c) **High-income organization for Economic Co-operation and Development (OECD) members**: This is a group of countries which works to build better policies for better lives.
(d) **Development Assistance Committee (DAC) members**: The member of this committee discuss issues surrounding development aid and poverty reduction in developing countries.

Table 4.2 Energy consumption (EC) per capita for different countries

S.No	Name of country	EC per capita (kWh), 2014	Total EC per capita (kWh), 2022	Percentage increase in EC (%)
(a) **Some of developed countries**: EC per capita > 5000kWh (2014) [3, 4] and GNI > 13,000 US$ [6]				
1	Australia	10,071	60,660	502.73
2	Canada	15,588	100,310	543.51
3	France	6,940	37,041	433.73
4	Germany	7,035	40,153	470.76
5	Iceland	53,832	167,175	210.55
6	Italy	5,002	26,936	438.50
8	Korea, Rep	10,497	63,865	508.41
9	New Zealand	9,013	48,414	437.16
10	Norway	23,000	98,879	330.00
11	Russia	6,603	53,895	716.22
12	Saudi Arabia	9,401	84,262	796.31
13	Singapore	8,845	162,192	1733.71
14	South Africa	4,198	22,959	446.90
15	Sweden	13,480	60,494	352.14
16	United Kingdom (UK)	5,130	28,211	449.92
17	United States	12,994	73,677	467.00
(b) **Some of developing countries:** EC per capita between 500 and 4000 kWh (2014) [3, 4] and GNI between 4000 to 12,000 US$ [6]				
1	Algeria	1,363	14,561	968.31
2	Argentina	3,075	19,352	529.33
3	Belarus	3,690	28,871	682.41
4	Brazil	2629	15,692	507.15
5	Chile	3,880	23,348	501.75
6	China	3,905	28,072	618.87
7	Cyprus	3,625	23,358	544.35
8	Georgia	2,694	18,547	588.46
9	India	805	6,438	699.75
10	Indonesia	1,084	7,753	615.22
11	Paraguay	1,552	18,398	1085.44
(c) **Lest developed countries:** EC per capita < 500kWh (2014) [3, 4] and GNI between 200 and 4000 US$ [6]				
1	Angola	312	3,430	999.35
2	Bangladesh	**320**	2,685	739.06
3	Congo Dem. Republic	109	403	269.72

(continued)

Table 4.2 (continued)

S.No	Name of country	EC per capita (kWh), 2014	Total EC per capita (kWh), 2022	Percentage increase in EC (%)
4	Congo Republic	**203**	4,735	2232.51
5	Ethiopia	**69**	944	1268.11
6	Ghana	**351**	3,294	838.46
7	Haiti	**39**	1,164	2884.61
8	Myanmar	215	3,130	1355.81

4.2.2 Developing Countries

The countries which have not achieved (i) a significant degree of industrialization with respect to their populations since World War Two and (ii) have a medium to low standard of living, Table 4.2b.

4.2.3 Under-Developing Countries

According to the World Bank, the countries which have a gross national income (GNI) per capita such as US$ 1,018, are shown in Table 4.2c.

Most of the conditions mentioned above can be only achieved by generating thermal and electrical power as per requirements of individual countries. Further, each group desires to jump in higher energy per capita and GNI group without thinking of its consequences on environment and climate. Prior to World War Two, the level of CO_2 in environment was 280 ppm. In the year 2021, CO_2 levels in the atmosphere/environment have been measured as 417, resulting in global climate change [7]. This increase can be explained as follows:

(i) During electrical power generation, fossil fuel is burned to run turbines connected with generators, and hence electrical efficiency (power factor) is always less than one. At the time of burning, there is emission of both CO_2 and thermal energy.
(ii) The emission of CO_2 increases in the atmosphere.
(iii) The rejection of thermal energy into the atmosphere raises global atmospheric temperatures. The rise of atmospheric temperatures above glaciers is causing them to melt; in turn, increasing the level of the sea.

The atmosphere on Earth is the same, so emitted CO_2 and thermal energy migrate from one region to another on the planet, irrespective of country.

From Tables 4.2 and 4.3, one can infer that developed countries are more responsible than other countries. In order to solve this issue, an agreement was made between all countries to adopt the Kyoto protocol treaty to limit greenhouse gas[GHG: water vapor (H_2O), carbon dioxide (CO_2), methane (CH_4), nitrous oxide (N_2O), and ozone (O_3)] emission into atmosphere to sustain climate. The carbon dioxide (CO_2) is the primary greenhouse gas emitted through human activities into atmosphere. This protocol countries [**Annexure-I/B:** 37 industrialized countries (list of some industrialized countries out of 37are mentioned in Table 4.2a)] were responsible for about at least 55 percent of total carbon dioxide emissions in 1990 [3].

Following measure in addition to points mentioned in Sect. 4.1 to reduce GHG emission was proposed:

(i) **Plantation of trees**: The planting of trees by all 55 signatory countries of Annexure-I which consume carbon dioxide (CO2) from the air of atmosphere for photosynthesis (Chap. 1)
(ii) **Clean Development Mechanism (CDM)**: It encourages developed countries (Annexure-I) to invest in various fossil fuel based energy consumed technology and infrastructure in less-developed countries [Tables 4.2b and c],
(iii) **Emissions trading**: It allows participating countries to buy and sell emissions rights and thereby placed an economic value on greenhouse gas (GHG) emissions into the atmosphere.

The Kyoto protocol was effective from February 2005,

Table 4.3 The top 10 countries by energy consumption per capita are relatively wealthy and heavily industrialized [4]

Country	Year of data	Energy consumption per capita (kWh)
Iceland	2020	167,175
Qatar	2020	165,044
Singapore	2020	162,192
Bahrain	2019	145,193
Trinidad and Tobago	2020	123,800
Brunei	2019	121,637
United Arab Emirates	2020	117,686
Canada	2020	100,310
Norway	2020	98,879
Kuwait	2020	98,021

4.3 Energy Conservation

Energy Conservation Day has Been Celebrated on December 14 Every Year Since 1991

Energy Conservation is the effort made by all mankind to reduce the consumption of fossil fuel based energy for power production for day to day activities. Various measures for reduction in fossil based energy have been discussed in Sect. 4.1 and 4.2 respectively. Energy conservation is a main part of the concept of ecosufficiency/friendly. It can result in:

(a) Increased environmental quality.
(b) National security.
(c) Personal financial security.
(d) Higher savings.
(e) Preventing future non-renewable energy resource depletion.

Table 4.2 also provides the increase in consumed energy (kWh) per capita from 2014 to 2022 for some countries in different categories, namely developed, developing and least developed countries. One can observe that every country desires to upgrade itself from a lower category to a higher category by increasing their energy consumption (kWh) per capita.

The following observations can be made from Table 4.2:

(i) Least developed countries namely Angola, Congo republic, Ethiopia, Haiti and Myanmar have increased their energy consumption more than 1000%.
(ii) Some developing countries, namely Algeria and Paraguay, have also increased their energy consumption by more than 1000%.
(iii) Even some developed countries, namely Australia, Canada, Korea, Rep. (South Korea), Russia, Saudi Arabia and Singapore have increased their energy consumption from 500% to more than 1000% because they want to become superpower countries.

This development has took place/occurred in last 8–9 years after the Tokyo protocol. Every country has justifications for their actions in increasing their energy consumption per capita. But at what cost?. The only answer is on cost of existence of wildlife including humans, ecology, the environment and climate, which are all parts of **Nature**.

4.4 Energy Audit

An energy audit is an important tool/method to find out:

(i) Potentials for electrical/thermal energy efficiency measures of equipments/tools.

(ii) Assessing their financial viability/implication to be carried out at different levels.
(iii) Identifies low cost energy saving opportunities.

The main objectives of the energy audit are as follows:

- To understand use and waste of electrical/thermal energy within the system, namely residential and commercial buildings.
- To find cost effective alternative measures for reduction in energy losses and improvement in overall performance.
- To evaluate cost–benefit analysis on the basis of electrical/thermal energy saving for indentifying which energy efficient system/equipment are best to implement.

Further, conducting regular energy audit of any system will encourage the following:

- To reduce carbon footprint and energy conservation/saving improvement opportunities.
- To help understand better energy usage.
- To identify safety concerns regarding maintenance of electrical systems, wiring, and ventilation etc. This will make your homes and businesses safer.
- It will increase a system's resale value.
- It will help evaluation of reduction of carbon monoxide production in the home, industries or any business.

There are basic four different phases of energy audit which are as follows [9]:

1. **First Phase**

   In this phase, a review of energy use on the basis of past fuel consumption in the home, industry or any business should be done. Also, the hourly, daily, monthly and yearly energy flow chart within the system along with a list of used equipment and their energy demands should be prepared. The more detailed the energy usage data, preferably on an hourly basis, the better the actual analysis will be. On the basis of the collected/generated hourly data, the auditor will be able to calculate the actual total energy demand for given scenarios, namely daily and monthly seasonal variation, and will be able to set each system component into comparison. With this preliminary analysis, an experienced auditor can estimate for efficiency improvements in the system or building to save electrical and thermal energy.
2. **Second phase**

   In this case, the site assessment for any kind of building is done. The various components namely nature of wind for ventilation, lighting, passive cooling/heating, are examined, and their performance data is collected. Such a process can vary largely in terms of effort.

3. **Third phase**

   In this phase, the detailed data analysis done, and this is the most complex part of an energy audit. It involves:

   (a) Technical analysis: It includes a simple spreadsheet energy balance where all input and output energy parameters are determined on the basis of hour, day, month and year, or can be achieved by designated software packages. This energy spreadsheet is used for cost analysis.
   (b) Cost analysis: It includes current energy costs, costs for implementation of energy efficiency measures, potential energy savings over time (life of energy efficiency components), and payback period, life cycle cost analysis, operation and maintenance (O&M), energy matrices, namely energy payback time, energy production factor and life cycle energy analysis, are considered. Methodologies for analyzing the collected data vary widely and are subject to the system or process to be assessed.

4. **Fourth phase: Audit Report**

   In this phase, a final comprehensive report is prepared on the basis of the findings of the first three phases, which includes (i) all recommended energy efficiency measures, (ii) different combinations of energy efficiency measures, and (iii) energy saves which leads to cost and energy analysis.

   Energy saving is more economical than installation of energy plants

   Based on energy audit of any system, one can estimate carbon mitigation which can be used to evaluate carbon credit.

4.5 Carbon Credit (Adopted from Tiwari and Mishra [10])

4.5.1 Embodied Energy

Embodied energy is defined as: "the quantity of energy required by all of the activities associated with a production process, including the relative proportions consumed in all activities upstream to the acquisition of natural resources and the share of energy used in making energy equipment and in other supporting functions i.e. direct energy plus indirect energy" [11].

Thus, the aim of any embodied energy analysis is to quantify the amount of energy used to manufacture a material, product, component, and element. This involves the assessment of the overall expenditure of energy required to extract the raw material, manufacture products and components, build and maintain the component element being assessed. A secondary aim is to establish the embodied energy required to construct and maintain the item, component or building over the whole life cycle.

Like operational energy, embodied energy is an indicator of the level of energy consumption. Reducing energy consumption through better design with higher efficiency has been a goal of designers for many years. But the embodied energy portion of this consumption has largely been ignored. There are several reasons for this omission, including no clear assessment methodology, lack of data, lack of understanding, and a common belief that the embodied energy portion of assets energy consumption is insignificant. However, over recent years, the methodologies for assessment of each component of any system have been improved.

To be able to quantify the energy embodied in the construction of an asset, the quantities of materials must first be estimated through a process of desegregation and decomposition to a level of detail which allows for the separation of components into their principal materials. Energy intensities/densities of each material can then be multiplied by the quantities of individual materials and the products aggregated to obtain the total for each material, element. In addition to the embodied energy value, other environmental indicators can also be calculated, such as CO_2 emissions.

An expression for evaluating embodied energy (E_{in}) of any system can be written as follows:

$$E_{in} = \sum_{i=1}^{n} \rho_i \times m_i \tag{4.1}$$

where ρ_i and m_i are energy density/intensity and mass of ith components of any energy system.

4.5.2 Energy Density

Energy densities/intensities (ρ_i) are derived from energy analysis studies from various national and international sources. Among the difficulties encountered in using a wide variety of sources to verify values is the need to clarify definitions of system boundaries or whether the values are in terms of primary energy or delivered energy. To obtain an accurate and reliable database of energy intensities for all materials used in water assets is an enormous task in itself. It is a necessity for detailed comparisons of materials. The main requirement of embodied energy calculations at the design stage is obtaining accurate and useable material quantities and then combining them with currently available energy intensity values [10].

There are several methods used to carry out an energy analysis, namely (a) Process analysis (b) Input and output analysis and (c) Hybrid analysis. Traditionally, input–output analyses have been used to derive the embodied energy intensities [10].

Energy density of different materials generally used to manufacture renewable energy technology has been given in Table 4.4.

Table 4.4 List of embodied energy coefficients

Material	MJ/kg	MJ/m^3
Aggregate, general	0.10	150
Virgin rock	0.04	63
River	0.02	36
Aluminium, virgin	191	515,700
Extruded	201	542,700
Extruded, anodised	227	612,900
Extruded, factory painted	218	588,600
Foil	204	550,800
Sheet	199	537,300
Aluminium, recycled	8.1	21,870
Extruded	17.3	46,710
Extruded, anodised	42.9	115,830
Extruded, factory painted	34.3	92,610
Foil	20.1	54,270
Sheet	14.8	39,960
Asphalt (paving)	3.4	7,140
Bitumen	44.1	45,420
Brass	62.0	519,560
Carpet	72.4	–
Felt underlay	18.6	–
Nylon	148	–
Polyester	53.7	–
Polyethylterepthalate (PET)	107	–
Polypropylene	95.4	–
Wool	106	–
Cement	7.8	15,210
Cement mortar	2.0	3,200
Fibre cement board	9.5	13,550
Soil–cement	0.42	819
Ceramic		–
Brick	2.5	5 ,170
Brick, glazed	7.2	14,760
Pipe	6.3	–
Tile	2.5	5,250
Concrete		–
Block	0.94	–
Brick	0.97	–
GRC	7.6	14,820

(continued)

Table 4.4 (continued)

Material	MJ/kg	MJ/m^3
Paver	1.2	–
Pre-cast	2 0.0	–
Ready mix, 17.5 MPa	1.0	2,350
30 MPa	1.3	3,180
40 MPa	1.6	3,890
Roofing tile	0.81	–
Copper	70.6	631,160
Earth, raw		
Adobe block, straw Stabilized	0.47	750
Adobe, bitumen stabilised	0.29	–
Adobe, cement stabilised	0.42	–
Rammed soil cement	0.80	–
Pressed block	0.42	–
Fabric		
Cotton	143	–
Polyester	53.7	–
Glass	–	–
Float	15.9	40,060
Toughened	26.2	66,020
Laminated	16.3	41,080
Tinted	14.9	375,450
Insulation	–	–
Cellulose	3.3	112
Fibreglass	30.3	970
Polyester	53.7	430
Polystyrene	117	2,340
Wool (recycled)	14.6	139
Lead	35.1	398,030
Linoleum	116	150,930
Paint	90.4	118 per litre
Solvent based	98.1	128 per litre
Water based	88.5	115 per litre
Paper	36.4	33,670
Building	25.5	–
Kraft	12.6	–
Recycled	23.4	–
Wall	36.4	–

(continued)

Table 4.4 (continued)

Material	MJ/kg	MJ/m^3
Plaster, gypsum	4.5	6,460
Plasterboard	6.1	5,890
Plastics	–	–
ABS	111	–
High density polyethelene (HDPE)	103	97,340
Low density polyethelene (LDPE)	103	91,800
Polyester	53.7	7,710
Polypropylene	64.0	57,600
Polystyrene, expanded	117	2,340
Polyurethane	74.0	44,400
PVC	70.0	93,620
Rubber	–	–
Natural latex	67.5	62,100
Synthetic	110	–
Sand	0.10	232
Sealants and adhesives	–	–
Phenol formaldehyde	87.0	–
Urea formaldehyde	78.2	–
Steel, recycled	10.1	37,210
Reinforcing, sections	8.9	–
Wire rod	12.5	–
Steel, virgin, general	32.0	251,200
Steel, galvanised	34.8	273,180
Steel, imported, structural	35.0	274,570
Stone, dimension	–	–
Local	0.79	1,890
Imported	6.8	1,890
Straw, baled	0.24	30.5
Timber, softwood	–	–
Air dried, roughsawn	0.3	165
Kiln dried, roughsawn	1.6	880
Air dried, dressed	1.16	638
Kiln dried, dressed	2.5	1,380
Mouldings, etc	3.1	1,710
Hardboard	24.2	13,310
MDF	11.9	8,330

(continued)

Table 4.4 (continued)

Material	MJ/kg	MJ/m^3
Glulam	4.6	2,530
Particle bd	8.0	–
Plywood	10.4	–
Shingles	9.0	–
Timber, hardwood	–	–
Air dried, roughsawn	0.50	388
Kiln dried, roughsawn	2.0	1,550
Vinyl flooring	79.1	105,990
Zinc	51.0	364,140
Galvanising, per kg steel	2.8	–

4.5.3 Embodied Energy of Photo-Voltaic Module

The calculation of embodied energy of PV module which goes through various high technological processes has been described below.

The total embodied energy required for making individual components of the PV module is difficult with their manufacturing energy needs to be evaluated. Energy requirement in different processes for production of PV module are given in Table 4.5.

The embodied energy of PV module for 1 m^2 [13, 14] for the following specifications are given in Table 4.6.

The above data has been generated by excluding the embodied energy of balance of system (BOS) of PV system.

The basis of the above data generated is due to following reasons:

(i) Reduction of mass of solar cell: It may be due to development of new materials. Case (iii) has been considered as 10% of case (i).
(ii) Cell processing energy: It is reduced by 75% of case (i) to case (iii).
(iii) Cell efficiency: It is increased by 4%.

Table 4.5 Energy requirement (Energy density) in different processes for production of PV module

Process	Energy requirement
Silicon purification and processing	
(a) Metallurgical grade silicon (MG-Si) production from silicon dioxide (quartz, sand)	20 kWh per kg of MG-Si
(b) Electronic grade silicon (EG-Si) production from MG-Si	100 kWh per kg of EG-Si
(c) Czochralski Silicon (Cz-Si) production from EG-Si	290 kWh per kg of EG-Si
Solar cell fabrication	120 kWh per m^2 of Silicon cell
PV Module assembly	190 kWh per m^2 of PV module
Roof top integrated PV system	200 kWh per m^2 of PV module

Table 4.6 The embodied energy of PV module for 1 m^2 at different processes

Processes	MG-Si (kWh)	EG-Si (kWh)	Eg-Si for Cz-Si (kWh)	Fabrication (kWh) Solar cell PV module	Total (kWh)
Case (i)	48	230	483	120 190	1071
Case (ii)	26.54	127.30	267.33	60.3 125.4	607
Case (iii)	4.80	23.0 0	48.30	90 95	261

(iv) Elimination of wafer trimming and packaging.
(v) Reduction of embodied energy of module assembly by 50%.

Case (ii) is based on the average value of case (i) and case (iii) respectively [13, 14]. Above results show that an embodied energy of a PV module is reduced significantly from 1071 to 261 kWh/m^2 due to various reason described above.

The PV module itself is called the system. Other components are called balance of system (BOS). It comprises wiring, electronic components, foundation, support structure, battery, installation, etc. For open field installation, the concrete, cement and steel are the main components used for foundation and frame, which requires maximum energy.

(a) **Total embodied energy of PV system for open fieldcondition (stand alone system)**

If the energy requirement for open field installation for balance of system (BOS) is 500 kWh/m^2 of panel, then total embodied energy for case (i) will be given by

Total embodied energy of PV system for open field condition (E_{in}) **=** 1071 + 500 = 1571 kWh/m^2

(b) **Total embodied energy of PV system for roof top condition**

For rooftop integrated PV system, the embodied energy requirement for balance of system (BOS) is reduced to 200 kWh/m^2 of PV module/panel due to absence of foundation and structure for frame then total embodied energy for case (i) will be given by

Total embodied energy of PV system for roof top condition (E_{in}) **=** 1071 + 200 = 1271 kWh/m^2

Further, the present embodied energy for both conditions can have lower value (say about 25–75% of its present value) in future due to development of new materials used for BOS.

4.5.4 Energy Matrices

In the recent past, developments in the design and manufacture of renewable energy technology have been very rapid. They are now predicted to become a major player to meet the energy need of human beings for good living, particularly in developing and under developing countries. The energy matrices are important for renewable technologies as their use makes no sense if the energy used in their manufacture is more than they can produce in their lifetime. The energy payback time is always one of the criteria used for comparing the viability of one renewable technology against another. For example, the energy analysis of a PV module was conducted by Hunt [15] and it was reported that the energy payback time (EPBT) of a PV module is 12 years. The results reported by Hunt[5] are also in general agreement with those of Kato et al. [16] for a crystalline silicon (c-Si) solar cell module. Aulich et al. [17] evaluated the EPBT for a crystalline silicon module and it was concluded that the EPBT is 8 years; in this case, plastic materials were used for encapsulation for the Siemens C process. The energy payback time (EPBT) for a crystalline silicon (c-Si) solar cell module under Indian climatic condition for annual peak load duration is about four years [18]. Lewis and Keoleian [19] predicted the energy pay back time (EPBT) for an amorphous silicon (a-Si) solar cell module with efficiency of 5% as 7.4 years for the climatic conditions of Detroit, USA; the EPBT gets reduced to 4.1 years with the increase in the efficiency of the module to 9%. Srinivas et al. [20] reported that the energy payback time for an amorphous silicon (a-Si) solar cell module reduces to 2.6 years after considering gross energy requirement (GER) and the hidden energy. Battisti and Corrado [21] investigated the energy payback time for a conventional multi-crystalline building integrated system, retrofitted on a tilted roof, located in Rome (Italy); with the yearly global insolation (solar intensity) on a horizontal plane was taken as 1530 kWh/m^{2}y. They concluded that the energy payback time (EPBT) gets reduced from 3.3 year to 2.8 year.

To reduce the energy payback time as mentioned in the case of PV module, the following considerations for other renewable energy technologies should be adopted:

(i) Increase of efficiency.
(ii) Use the cost effective materials with low energy densities for longer life.
(iii) With minimum annual maintenance.

(a) **Energy pay back time (EPBT)**

The EPBT depends on (i) the energy spent to prepare the materials used for fabrication of the system and its components (this is referred as embodied energy) and (ii) the annual energy yield (output) obtained from such system. To evaluate embodied energy of various components of system, the energy densities of different materials are required. This will be discussed in detail in the next section. **The energy pay back time (EPBT) is the total time period required to recover the total energy spent to prepare the materials (embodied energy) used for fabrication renewable energy technologies.** It is the ratio of embodied energy and the annual energy output from the system. It can be mathematically can be expressed as

$$EPBT\ in\ years = \frac{Embodied\ energy\ (E_{in})}{Ennual\ energy\ output\ (E_{aout})} \tag{4.2}$$

The numerical values of EPBT should be as minimum as for as possible to make the renewable energy technologies cost effective to users.

(b) **Energy production factor (EPF)**

It is used to predict the overall performance of the system. It is defined as the ratio of the output energy and the input energy or it can also be expressed as the inverse of EPBT.

(i) **On annual basis**

$$\chi_a = \frac{E_{aout}}{E_{in}} \tag{4.3a}$$

or,

$$\chi_a = \frac{1}{EPBT},\ years^{-1} \tag{4.3b}$$

If $\chi_a \to 1$, for $EPBT \to 1$ the system is worthwhile otherwise it is not worth from energy point of view.

(ii) **On basis of lifetime**

$$\chi_a = \frac{T}{EPBT} = T\frac{E_{aout}}{E_{in}} (dimensionless) \tag{4.3c}$$

For a given life of system, T, χ_a should be more than one. The value of χ_a should be as maximum as for as possible of cost effectiveness of system.

(c) **Life cycle conversion efficiency (LCCE)**

The LCCE is the net energy productivity of the system with respect to the solar input (radiation) over the life time of the system (T years). The expression for LCCE is given by

$$[\phi(t)] = \frac{E_{aout} \times T - E_{in}}{E_{a,sol} \times T}\ (in\ fraction) \tag{4.4a}$$

$$[\phi(t)] = \frac{E_{aout} \times T - E_{in}}{E_{a,sol} \times T} \times 100\ (in\ \%) \tag{4.4b}$$

The numerical value of LCCE is always less than one. However if the value of LCCE approaches one, it is the best technology from energy point of view.

(c) **Annual electrical output** (E_{aout})

One can observe that embodied energy (E_{in}), annual electrical output (E_{aout}) and life of renewable energy technology (T) have very important role to evaluate the energy matrices (Eqs. 4.2–4.4).

$$\text{Annual output of PV module } (E_{aout}) = \eta_m \times \bar{I} \times A_m \times N \times n = \eta_m \times E_{a,sol} \times A_m \quad (4.5)$$

where $\boldsymbol{E_{a,sol}} = \bar{\boldsymbol{I}} \times N \times n$, $\bar{\boldsymbol{I}}$, η_m and $A_m\eta_0$ are annual solar radiation, daily average solar radiation, electrical efficiency of PV module and area of PV module. N and n are the number of sunshine hours and clear days in a year. The is the annual average values of solar intensity which varies from place to place, e.g. its value for Port Hedland (NW Australia, Sydney and India are 2494, 1926 and 1800 kWh/m^2/year respectively. This indicates that annual electrical out put for Port Hedland (NW Australia) will be maximum.

For the following parameters namely solar cell electrical efficiency (η_m) = 0.12, A_m = 1m^2 and

(a) For Port Hedland (NW Australia) climatic condition $\left(E_{a,sol} = 2494\,\text{kWh/m}^2\right)$,
Annual output of PV module for 1m^2 (E_{aout}) = 0.12 × 2494 × 1 = 299.28 kWh
(b) For Sydney climaticcondition$\left(E_{a,sol} = 1926\text{kWh/m}^2\right)$ Annual output of PV module for 1m^2 (E_{aout}) **=** 0.12 × 1926 × 1 = 231.12 kWh, and similarly.

(c) For Indian $\left(E_{a,sol} = 1800\text{kWh/m}^2\right)$ climatic condition for the same design parameters,
Annual output of PV module for 1m^2 (E_{aout}) =0.12 × 1800 × 1 = 216 kWh

If 1kWh power generation is equivalent to 0.98 kg of CO_2 emission by thermal power plant than CO_2 mitigation can be obtained from the following expression by using renewable energy sources for electrical energy generation as

$$CO_2\text{mitigation in kg due to coal} = E_{aout}(kWh) \times 0.98\left(\frac{Kg}{kWh}\right) \quad (4.6a)$$

Referring to Table 4.1, for other fossil fuel,

$$CO_2\text{mitigation in kg due to petroleum} = E_{aout}(kWh) \times (0.8\ to\ 0.93)\left(\frac{kg}{kWh}\right) \quad (4.6b)$$

$$CO_2\text{mitigation in kg due to natural gas} = E_{aout}(kWh) \times (0.443)\left(\frac{Kg}{kWh}\right) \quad (4.6c)$$

If the life of a PV module is considered to be T = 30 years, then the matrices of a PV module can be obtained from Eqs. (4.2–4.4). The results for three cases are given in Table 4.7 [13]. The three cases are as follows:

Case (i): In this case, embodied energy of PV module has been considered to be 1071 kWh which is based the energy consumed by different processes.
Case (iii): In this case, if energy consumption becomes 10% of each process due to advanced technology development, embodied energy for case (iii) is to be as 261 kWh.
Case (ii): In this case, an average value of case (i) and case (iii) for each processes has been considered. The embodied energy in this case is 607 kWh.

From Table 4.7, shows that PV modules are best suited for Port Hedland (NW Australia) due to the lowest value of EPB and the highest value of EPF and LCCE. This is because of higher values of solar radiation in comparison with Sydney and India.

Example 4.1 Evaluate life cycle conversion efficiency (LCCE) for Indian climatic conditions for life of a PV system of 10 and 20 years for one meter square with balance of system (BOS) on the roof top.

Solution:

From Eq. 4.4, one has an expression for LCCE as follows:

$$\phi(t) = \frac{E_{aout} \times T - E_{in}}{E_{a,sol} \times T}$$

Here, $E_{in} = 1271$ kWh/m^2 (Sect. Embodied Energy of Photo-Voltaic Module), (E_{aout}) $= 216$ kWh, $E_{a,sol} = 1800\text{kWh/m}^2$
Substitute the above value in given equation, one gets

$$[\phi(t)]_{10} = \frac{216 \times 10 - 1271}{1800 \times 10} \times 100 = 4.93\%$$

$$[\phi(t)]_{20} = \frac{216 \times 20 - 1271}{1800 \times 20} \times 100 = 8.46\%$$

Here, it important to note that the life of PV system/module plays an important role.

Table 4.7 The energy matrices in different countries for three cases without balance of system (BOS)

City	Port Hedland			Sydney			India		
	EPBT (Years)	EPF	LCCE (Fraction)	EPBT (Years)	EPF	LCCE (Fraction)	EPBT (Years)	EPF	LCCE (Fraction)
Case (i)	3.60	8.30	0.105	4.64	6.47	0.101	4.96	6.05	0.100
Case (ii)	2.00	15.0	0.112	2.62	11.45	0.109	4.96	6.05	0.100
Case (iii)	0.87	34.5	0.116	1.13	26.55	0.115	2.81	10.68	0.109

Example 4.2 Repeat Example 4.1 without balance of system.
Solution:

Without balance of system, $E_{in} = 1071\, kWh/m^2$

From Eq. 4.4, one has an expression for LCCE as follows:

$$\phi(t) = \frac{E_{aout} \times T - E_{in}}{E_{a,sol} \times T}$$

Here, $E_{in} = 1271$ kWh/m^2 (Sect. 4.5.3), (E_{aout})=216 kWh, $E_{a,sol} = 1800\,\text{kWh/m}^2$

Substitute the above value in given equation, one gets

$$[\phi(t)]_{10} = \frac{216 \times 10 - 1071}{1800 \times 10} \times 100 = 6.05\%$$

$$[\phi(t)]_{20} = \frac{216 \times 20 - 1071}{1800 \times 20} \times 100 = 9.05\%$$

From Examples 4.1 and 4.2, it is further noticed that balance of system (BOS) plays an important role in the life cycle energy efficiency.

4.5.5 *Earned Carbon Credit [22]*

The trading of carbon credits was therefore created to curb the effects of greenhouse gases by reducing the carbon footprint. Carbon credits are defined as "a key component of national and international emissions trading schemes that have been implemented to mitigate global warming". They provide a way to reduce greenhouse effect emissions on an industrial scale by capping the total annual emissions and letting the market assign a monetary value to any shortfall through trading. Credits can be exchanged between businesses or bought and sold in international markets at the prevailing market price. Credits can be used to finance carbon reduction schemes between trading partners around the world. There are also many companies that sell carbon credits to commercial and individual customers who are interested in lowering their carbon footprint on a voluntary basis. These carbon offsetters purchase the credits from an investment fund or a carbon development company that has aggregated the credits from individual projects. The quality of the credits is based in part on the validation process and sophistication of the fund or development company that acted as the sponsor to the carbon project.

4.5.5.1 Formulation

If unit power is used by a consumer and the losses due to poor domestic appliances is La, then the transmitted power should be $\frac{1}{1-L_a}$ units. If the transmission and distribution losses are L_{td}, then the power that has to be generated in the power plant is $\frac{1}{1-L_a} \times \frac{1}{1-L_{td}}$ units.

The average CO_2 equivalent intensity for electricity generation from coal is approximately 0.98 kg of CO_2 per kWh at the source. Thus, for unit power consumption by the consumer, the amount of CO_2 emission is $\frac{1}{1-L_a} \times \frac{1}{1-L_{td}} \times 0.98$

The annual CO_2 emission can be expressed as:

$$CO_2 \text{ emission per year} = \frac{E_{in}}{T} \times \frac{1}{1 - L_a} \times \frac{1}{1 - L_{td}} \times 0.98\, kg \tag{4.7}$$

where E_{in} is the embodied energy input and T_s is the lifetime of the system.

The CO_2 emission over the lifetime of the system is:

$$(CO_2)_{Emission} = E_{in} \times \frac{1}{1 - L_a} \times \frac{1}{1 - L_{td}} \times 0.98\,\text{kg} \tag{4.8}$$

The net CO_2 mitigation over the lifetime of the system with the help of Eq. 4.6 in kg can be written as follows:

$$\begin{aligned}\text{Net } CO_2 \text{ mitigation(kg)} &= \text{Total } CO_2 \text{ mitigation} - \text{total } CO_2 \text{ emission} \\ &= (E_{aout} \times T \times 0.98) - (CO_2)_{Emission};\ \text{kg}\end{aligned} \tag{4.9}$$

where E_{aout} is the overall electrical/exergy gain.

The net CO_2 mitigation over the lifetime of the system in tones can be written as follows:

$$\text{Net } CO_2 \text{ mitigation (tones)} = \left[(E_{aout} \times T \times 0.98) - (CO_2)_{Emission} \times 10^{-3}\right] \tag{4.10}$$

If CO_2 emission is being traded at US$ cost per tones of CO_2 mitigation, then the carbon credit earned by the system for different fossil fuel (Table 4.1) is given by

$$\begin{aligned}&\text{Net carbon } (CO_2) \text{ credit earned (US\$) due to coal} \\ &= \left[[E_{aout} \times T \times (0.980\boldsymbol{r}1.001)] - (CO_2)_{Emission} \times 10^{-3}\right] \\ &\times \text{US\$ cost per tones}\end{aligned} \tag{4.12a}$$

$$\begin{aligned}&\text{Net carbon } (CO_2) \text{ credit earned (US\$) due to petroleum} \\ &= \left[(E_{aout} \times T \times (0.8\, to\, 0.93)) - (CO_2)_{Emission} \times 10^{-3}\right] \\ &\times \text{US\$ cost per tones}\end{aligned} \tag{4.12b}$$

$$\begin{aligned}&\text{Net carbon } (CO_2) \text{ credit earned (US\$) due to natural gas} \\ &= \left[(E_{aout} \times T \times 0.443) - (CO_2)_{Emission} \times 10^{-3}\right] \\ &\times \text{US\$ cost per tones}\end{aligned} \tag{4.12c}$$

4.5.5.2 A Case Study with the BIPVT System

If unit power is used by a consumer and the losses due to poor domestic appliances are around 20%, then the transmitted power should be $\frac{1}{1-0.2} = 1.25$ units. If the transmission and distribution losses are 40%, common in Indian conditions, then the power that has to be generated in the power plant is $\frac{1.25}{1-0.4} = 2.08$ units. The average CO_2 equivalent intensity for electricity generation from coal is approximately 0.98 kg of CO_2 per kWh at the source. Thus, for unit power consumption by the consumer the amount of CO_2 emission is $2.08 \times 0.98 = 2.04$ kg. For the BIPVT system, the annual net CO_2 mitigation in tones of CO_2 is given by:

$$\text{Annual net } CO_2 \text{ mitigation in tones} = (E_{aout} - \frac{E_{in}}{T}) \times 2.04 \times 10^{-3} \quad (4.13)$$

The data of annual energy in kWh and embodied energy (kWh) with lifetime (T) of different solar cell material of Table 4.8 has been used in Eq. 4.13 to determine annual net CO_2 mitigation and carbon credit earned and given in same Table 4.8. One can see that maximum carbon credit earned is from Mono-crystalline silicon (c-Si) is US$1540 followed by Multi-crystalline silicon (mc-Si) as US$1361 due to long-life durability and its electrical efficiency.

Table 4.8 The data of lifetime, embodied energy, annual energy, annual CO_2 mitigation and carbon credit earned by different solar cell material for roof area of 45 m^2

S.No	Solar cell materials	Life (T), years	Embodied energy (E_{in}), MJ kWh	Overall annual exergy, kW	Overall annual exergy for N = 5 h kWh (E_{aout})	Annual CO_2 mitigation, tonnes	Carbon credit earned, US$
1	Mono-crystalline silicon (c-Si)	30	607,613 168,781	16,224	81,120	154	1,540
2	Multi-crystalline silicon (mc-Si)	30	540,628 150,174	14,352	71,760	136.1	1,361
3	Ribbon silicon (r–Si)	25	409,716 113,819	12,512	62,560	118.3	1,183
4	Amorphous silicon (a-Si)	20	272,324 75,645	7,790	38,950	71.74	717.4
5	Cadmium telluride (CdTe,)	15	211,984 58,884	9,547	47,735	89.37	893.7
6	Copper indium gallium selenide (CIGS)	5	63,937 17,760	11,037	55,185	54.97	1,099

Note The CO_2 emissions are being traded at $US 10 per tonne of CO_2 mitigation and sunshine hour has been considered as 5 h

Example 4.3 Evaluate annual CO_2 emission for PV system/module without balance of system for all cases mentioned in Table 4.6 for Indian climatic condition for T = 20 years.

Solution:

From Eq. (4.13) we have

$$\text{Annual net } CO_2 \text{ mitigation in tones} = (E_{aout} - \frac{E_{in}}{T}) \times 2.04 \times 10^{-3}$$

From Table 4.6, we have the following embodied energy (E_{in}) and annual electrical energy (E_{aout})

Case (i): $E_{in} = 1071kWh$, and E_{aout} = 216 kWh (Example 4.2)

Case (ii): $E_{in} = 607kWh$ and E_{aout} = 216 kWh (Example 4.2)

Case (iii): $E_{in} = 261kWh$ and E_{aout} = 216 kWh (Example 4.2)

Now,

Case (i): Annual CO_2 mitigation $= \left[216 - \frac{1071}{20}\right] \times 2.04 \times 10^{-3} = 0.331$ tonnes

Case (ii): Annual CO_2 mitigation $= \left[216 - \frac{607}{20}\right] \times 2.04 \times 10^{-3} = 0.378$ tonnes

Case (iii): Annual CO_2 mitigation $= \left[216 - \frac{261}{20}\right] \times 2.04 \times 10^{-3} = 0.414$ tonnes

This shows that annual CO_2 mitigation is increased by decreasing embodied energy.

Problems

Problem 4.1 Prepare another table similar to Table 4.7 with embodied energy for stand alone and roof top.

Hint: See Sect. Embodied Energy of Photo-Voltaic Module

Problem 4.2 Evaluate energy matrices for case (ii) and case (iii) of Table 4.6

Hint: See Table 4.7 calculation

Problem 4.3 Evaluate annual CO_2 emission for PV system/module with balance of system for open field condition (stand alone) and roof top for all cases mentioned in Table 4.6 for Indian climatic condition for T = 20 years.

Hint: Add 500 kWh for field condition and 200 kWh for roof top in embodied energy of Table 4.5 and see Example 4.3

Problem 4.4 Prepare the Table 4.8 for other fossil fuel given in Table 4.1

Hint: Repeat the calculation similar to Table 4.8

Objective Questions

4.1 The life cycle conversion efficiency (LCCE) will be zero for

(a) $E_{aout} \times T > E_{in}$
(b) $E_{aout} \times T < E_{in}$
(c) $E_{aout} \times T = E_{in}$
(d) None of these

Answer: (c)

4.2 The life cycle conversion efficiency (LCCE) will be positive for installation of PV system

(a) $E_{aout} \times T > E_{in}$
(b) $E_{aout} \times T < E_{in}$
(c) $E_{aout} \times T = E_{in}$
(d) None of these

Answer: (a)

4.3 The energy payback time (EPBT)

(a) Decreases with increase of annual electrical energy
(b) Increases with increase of annual electrical energy
(c) Inversely proportional to annual electrical energy
(d) None of these

Answer: (a) and (c)

4.4 Energy consumption per capita (kWh/person) is maximum for

(a) Developed countries
(b) Developing countries
(c) Under developing countries
(d) All of these

Answer: (a)

4.5 Energy consumption per capita (kWh/person) is maximum for

(a) Developed countries
(b) Developing countries
(c) Under developing countries
(d) All of these

Answer: (c)

4.6 Energy conservation day is celebrated in each year on

(a) January 14
(b) March 14
(c) June 14
(d) December 14

Answer: (d)

4.7 Kyoto protocol has been effective since

(a) 2020
(b) 2010
(c) 2015
(d) 2005

Answer: (d)

4.8 Kyoto protocol treaty means

(a) Greenhouse gas emission (GHG)
(b) Maximum use of renewable energy sources
(c) CO2 mitigation
(d) All of these

Answer: (d)

4.9 CO_2 emission in environment since WW-II is due to

(a) Industrialization
(b) Maximum use of fossil fuel
(c) Minimum use of renewable sources
(d) All of these

Answer: (d)

4.19 Annual CO2 mitigation will be positive if

(a) $E_{aout} > \frac{E_{in}}{T}$
(b) $E_{aout} < \frac{E_{in}}{T}$
(c) $E_{aout} = \frac{E_{in}}{T}$
(d) In all cases

Answer: (a)

4.20 Which fuel is most preferred for power generation?

(a) Hard coal
(b) Petroleum
(c) Natural gas
(d) All of these

Answer: (c)

References

1. Energy Conservation: Definition, Ways to Conserve and … https://www.embibe.com/exams/energy-conservation
2. Specific carbon dioxide emissions of various fuels https://www.volker-quaschning.de › CO2-spez › index_e
3. Energy Consumption by Country 2022 https://worldpopulationreview.com › ...
4. Electric power consumption (kWh per capita) - World Bank Data, 2014 https://data.worldbank.org › EG.USE.ELEC.KH.PC
5. Mapped: Energy Consumption Per Capita Around the World https://elements.visualcapitalist.com › ...
6. List of countries by GNI (nominal) per capita – Wikipedia https://en.wikipedia.org ›
7. How have CO2 levels changed since the industrial revolution? https://www.weforum.org ›
8. Kyoto Protocol | History, Provisions, & Facts | Britannica https://www.britannica.com ›
9. Energy Audit: What and Why Important? https://justenergy.com ›
10. G. N. Tiwari and R.K. Mishra, Advanced Renewable Energy Sources, Royal Society of Chemistry (RSC), (UK), 2011
11. G.J. Treloar, *Energy analysis of the construction of office buildings*, Master of Architecture Thesis, Deakin University, Geelong, 1994.
12. I. Boustead and G.F. Hancock, *Handbook of Industrial Energy Analysis*, Ellis Horwood Publishers, 1979, 309–410
13. G.N. Tiwari, M.K. Ghosal, *Renewable Energy Resources: Basic Principles and Applications* (Narosa Publishing House, New Delhi, India, 2005)
14. G. N. Tiwari and Swapnil Dubey, Fundamentals of Photovoltaic Modules and Their Applications, Royal Society of Chemistry (RSC), (UK), 2010
15. L.P. Hunt, *IEEE PV Specialists Conference* (Piscataway, NJ, 1986), pp.347–352
16. K. Kato, A. Murata, K. Sakuta, Progress in Photovoltaic Research Applications **6**(2), 105–115 (1998)
17. H.A. Aulich, F.W. Schulz, B. Strake, *IEEE PV Specialist Conference* (Piscataway, NJ, 1986), pp.1213–1218
18. R. Prakash, N.K. Bansal, Energy Sources **17**, 605–613 (1995)
19. G. Lewis, G. Keoleian, *National Pollution Prevention Center* (University of Michigan, School of Natural Resources and Environment, 1996)
20. K.S. Srinivas, M. Vuknic, A.V. Shah, R. Tscharner, *6th International Photovoltaic Science and Engineering Conference (PVSEC-6)* (New Delhi, India, 1992), pp.403–413
21. R. Battisti, A. Corrado, Energy **30**, 952–967 (2005)
22. B. Agrawal, G.N. Tiwari, *Building Integrated Photovoltaic Thermal Systems* (RSC Publishing, U.K., 2010)

Chapter 5
Climate Change

5.1 Introduction

Climate change means global warming. It is due to fact that the average global air temperature on the planet Earth increases in abnormal trends in the timing of seasons. It also happened beyond natural variations of ambient air temperatures on land and in the ocean. The present rise in global average temperature is faster than previous changes prior to Word War Two due to significant increases in energy consumption (kWh) per capita by various countries (Table 4.2). The average global air temperature has been increasing (see Fig. 5.1) due to the following events [1, 2]:

(a) Burning biomass fossil fuels for domestic applications, fossil fuel (coal, oil, or gas), deforestation etc. to generate electricity and heat.
(b) Agricultural and industrial growth practices increase the greenhouse gases (GHG) namely carbon dioxide (CO_2) and methane (CH_4).
(c) Heat waves and wildfires (intensified by heat and drought) are becoming more common, expanding desert land.
(d) Greenhouse gases absorb some of the heat that the Earth emits as long wavelength radiation after absorption of solar radiation.

Larger amounts of these gases trap more heat in Earth's lower atmosphere causing global warming. Burning of fossil fuels (coal, oil and gas) is the main contributor to global climate change. It accounts for over 75% of global greenhouse gas emissions and nearly 90% of all carbon dioxide (CO_2) emissions into the atmosphere.

Due to global warming/climate change, the following effects have been observed:

(a) Changing weather patterns and disrupting the usual balance of nature.
(b) Deserts are expanding rapidly, affecting agricultural yield.
(c) Increased warming in the Arctic has contributed to melting permafrost, glacial retreat and sea ice loss.
(d) More intense storms, droughts, and other weather extremes.
(e) Rapid environmental change in mountains and coral reefs.

G. N. Tiwari, *Energy, Ecology and Environment*,
https://doi.org/10.1007/978-981-99-3997-8_5

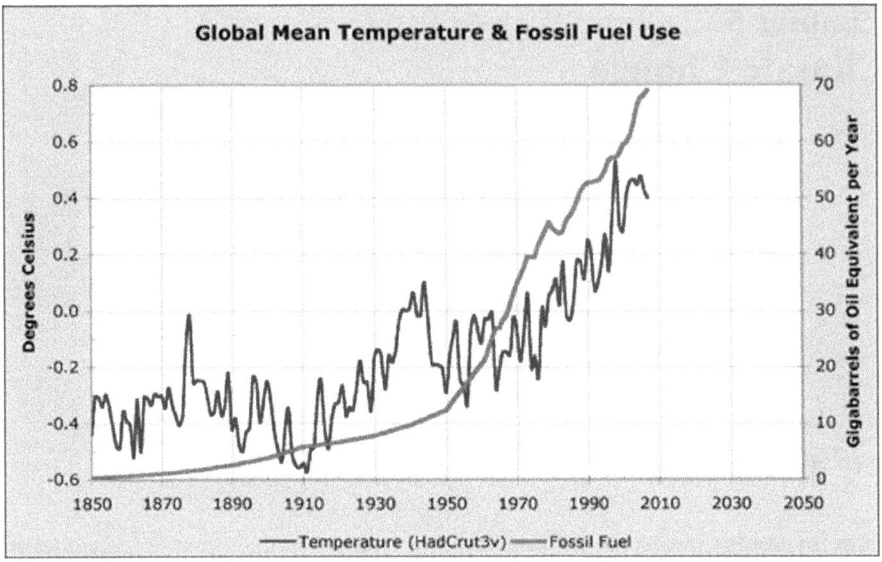

Fig. 5.1 How the global temperature and fossil fuels relate and increase at similar rates [1]

(f) Many species in the Arctic are being forced to relocate, or becoming extinct.
(g) Human/people are increasingly threatened with food and water scarcity.
(h) Increased flooding and heat, more disease, and economic loss.
(i) Human migration and conflict more common.
(j) Global human health facing its greatest challenges of the century.
(k) Sea level rising due to disappearing glaciers and melting of the Greenland ice sheet, and

The World Health Organization (WHO) calls this the greatest threat to global health of all living organisms, especially human beings, in the twenty-first century due to climate change.

At the Paris Agreement, which took place in the year 2015, all nations collectively agreed to keep warming level of Earth's surface temperature "well under 2 °C".

Even if efforts made by various countries to minimize future warming are partially successful, these efforts must continue for centuries to avoid (i) ocean heating, (ii) ocean acidification and (iii) sea level rise.

Many of these impacts are as follows:

(a) It is evident at the current 1.2 °C level of global warming and
(b) Climate change/global warming would still reach 2.7 °C by the end of the century.

To reduce half of CO_2 emissions by 2030, global warming should be limited to 1.5 °C.

Further, achieving the net-zero CO_2 emissions is possible by 2050 [3].

5.2 Reasons for Climate Change

There are many reasons for climate change. However, some main reasons of climate changes will be discussed in the following sections.

5.2.1 Wildfire

A wildfire is an unplanned fire which leads to a deterioration of the air quality, and loss of property, crops, resources, animals and people (see Fig. 5.2). It happens in dry conditions (drought) during high winds. Wildfire burns in a natural areas such as a forests, grasslands/prairies, grass homes in rural areas etc. Wildfires are generally caused by either human error or a natural phenomenon (lightning). It can happen at any time or anywhere throughout the world. Wildfires can disrupt (i) transportation, (ii) communications, (iii) power and gas services and (iv) water supply.

Fig. 5.2 Wildfire

5.2.2 The Anthropogenic Flux

Carbon emission due to industrialization (Sect. 2.3.3, Fig. 2.2b) since World War Two. It has happened due to increases of the level of comfort of humans.

5.2.3 The droughts

A drought can be defined as "a deficiency of precipitation during raining seasons over an extended period of days, month and years which results in a water shortage inside the ground." Drought, see Fig. 5.3, can be felt by judging the level of precipitation, ambient air temperature, stream waves, ground reservoir water levels, soil moisture, and snowpack etc. Due to drought, there can be imbalances in the ecosystem, agriculture, economy, chances of wildfire, quick water evaporation and climate change etc.

5.2.3.1 Effect of Drought

The Drought can effect the following:

(i) Environment, climate and economy
(ii) Drying of wetlands
(iii) Loss of biodiversity
(iv) Water scarcity

Fig. 5.3 Drought effected soil/ground

(v) Energy sector
(vi) Heat waves
(vii) Mass migration and humanitarian crisis
(viii) Agriculture

5.2.4 *Bleaching of Coral and Ocean Acidification [5, 6]*

Corals are sedentary (stagnant) animals which grow in colonies of many individual 'polyps' (jantu). Bleaching happens if coral polyps (jantu) eject algae (zooxanthellae) which live inside their tissue to cause the coral to be converted into white color as shown in Fig. 5.4. The reason for this process are many including (i) various environmental stressors, (ii) changes in ocean temperature, (iii) light, or (iv) nutrients. Algae (zooxanthellae) processes are photosynthetic, in the presence of light (solar radiation) in stagnant sea water (see Sect. 1.5) with rises of sea water temperature. In this process, they also begin to produce reactive oxygen species. This is toxic to the coral and hence it expels the zooxanthellae (algae) and produces coral coloration with transparency. It reveals the coral skeleton made of calcium carbonate ($CaCO_3$). Some coral may be blue, yellow, or pink due to pigment proteins in the coral. The overall warming of global sea water, even by 0.5 °C (El Niño event: a little boy), can led to bleaching at a large scale at coral reefs around the whole world.

Greenhouse gas (GHG) emissions generated by both natural and human means are responsible for ocean acidification. It is another major threat to the state of the world's oceans which cover about 70% of Earth. Many human activities like burning fossil fuels for their need release carbon dioxide (CO_2) into the atmosphere, and

Fig. 5.4 Photograph of bleaching coral

oceans absorb some of it, and it and dissolves in sea water which forms a mild acid. It changes the overall acidity of the sea water which is measured in pH scale ranging from 0 (fully acidic) to 14 (fully alkaline) through 7 (neutral). This is referred to as ocean acidification. Meanwhile, ocean acidification threatens sea life including corals, shellfish (mollusks and crustaceans), some algae, sea urchins and starfish. It varies between regions, species and ecosystems.

Both bleaching coral and ocean acidification pose threats to human commercial food supplies, such as snappers and groupers. They live and breed among coral reefs. They are threatened by bleaching.

5.2.4.1 Solution of Bleaching and Acidification

Greenhouse gas emissions (GHG) are the main reason for ocean acidification which increases sea water temperature. The increase of sea water cause coral bleaching as mentioned above. Any efforts to minimize green house gas (GHG) emissions will bring benefits both on land and at sea level. However, global greenhouse gas emissions have risen steadily in recent years particularly after WW-II.

There is so far less chance to avert an increase in ocean water acidity and temperature without a drastic reduction in global greenhouse gas (GHG) emissions.

5.3 Reducing Greenhouse Gas (GHG) Emission

The GHG emissions in atmosphere as per the Paris agreement require the following:

(a) Switching away from using burning fossil fuels.
(b) Using electricity generated from low-carbon sources.
(c) Phasing out coal-fired power plants.
(d) Increasing use of wind, solar, and other types of renewable energy.
(e) Reduction of energy use.
(f) Removal of carbon from the atmosphere by increasing forest cover [one acre of boreal forest (found in Canada, Alaska, Northern Europe, and Siberia) absorbs an average of 1.3 tons of carbon dioxide annually, one acre of tropical forest can absorb an average of 9.8 tons annually, and one acre of temperate deciduous forest, can absorb an average of 4.9 tons per acre annually; see Carbon Footprint: A Study of Unit and Dimensions. http://dimacs.rutgers.edu/mpe/cabon
(g) Farming that capture carbon in soil [4].
(h) Better coastline protection.

5.3.1 Concentrations of Various Greenhouse Gases in Atmosphere [7]

The yearly variation in concentration of carbon dioxide (CO_2), Methane (CH_4), Nitrogen oxide (N_2O), halogenated gases and ozone (O_3) has been shown in Fig. 5.5.

Figure 5.5a shows concentrations of carbon dioxide (CO_2) in the atmosphere from hundreds of thousands of years ago (before the common era, BCE) to the year 2021 (current era, CE).This figure shows that the concentration in ppm (parts per million) of carbon dioxide (CO_2) concentrations have increased significantly from 280 ppm prior to the beginning of the industrial era (World War Two) to 414 ppm in the year 2021. This is nearly 47.86% $\approx$ 48% due to human activities to meet their electrical energy demands.

The concentrations of methane (CH_4) in ppb (parts per billion) in the atmosphere have been shown in Fig. 5.5b. This variation is between hundreds of thousands of years ago to 2021. There is an increase of about 50% of methane (CH_4) concentration between from 1950 to 2020. This increase is in concentration of methane in atmosphere is due to human agriculture activities and fossil fuel use.

The yearly variation of nitrogen oxide (N_2O) during the past thousand years till the year 2021 has been shown in Fig. 5.5c. The 19.28% $\left[\frac{334-280}{280} \times 100\right]$ variation in concentration of nitrogen oxide (N_2O) in atmosphere between the years 1950 to 2021 has been observed due to only agricultural activities.

The yearly increase concentration measured in parts per trillion (ppt) of fluorine (F_2), chlorine (Cl_2), or bromine (BR_2) (halogenated gases) in the atmosphere between the year 1975 and 2021 has been shown in Fig. 5.5d with the scale increases by a factor of 10. From the figure,one can observe that the concentrations of many of some halogenated gases are zero a few decades ago. However, it is increased significantly due to industrial products and processes since WW-II which is responsible for Earth's protective ozone layer depletion. Only because of this, some chemicals factories have been phased out.

The variation in decrease of an average amount of ozone (O_3) layer in the Earth's atmosphere per year has been given in Fig. 5.5e. The effects of ozone depend on its altitude which is located vertically in the atmosphere. This decrease in thickness of, ozone (O_3) layer in Dobson unit is based on satellite measurements. The total column ozone which is called layer which is the "thickness" or density of ozone throughout all layers of the Earth's atmosphere is measured in Dobson units (in unit of 10 μm). This ozone layer is divided between the troposphere and the stratosphere, Fig. 1.5. One can conclude that the total amount of ozone (O_3) layer in stratosphere of the atmosphere (approximately 9.6 km to 48 km above the Earth's surface) is decreased by more than 4% between 1979 and 2020. Because of this decrease in ozone (O_3) layer in stratosphere, there is health issue in all living organism including human Sect. 1.4).

Due to presence of water vapor (H_2O) in abundant greenhouse gas in the atmosphere, there is a small direct influence on atmospheric concentrations of water

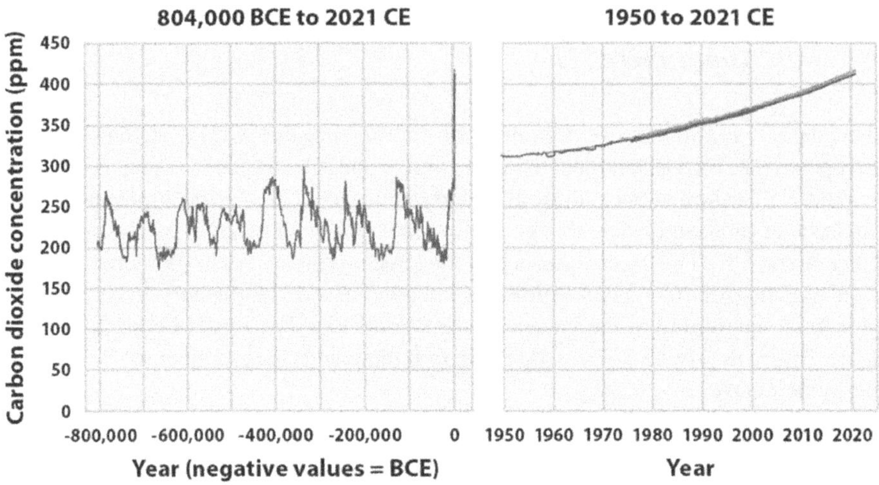

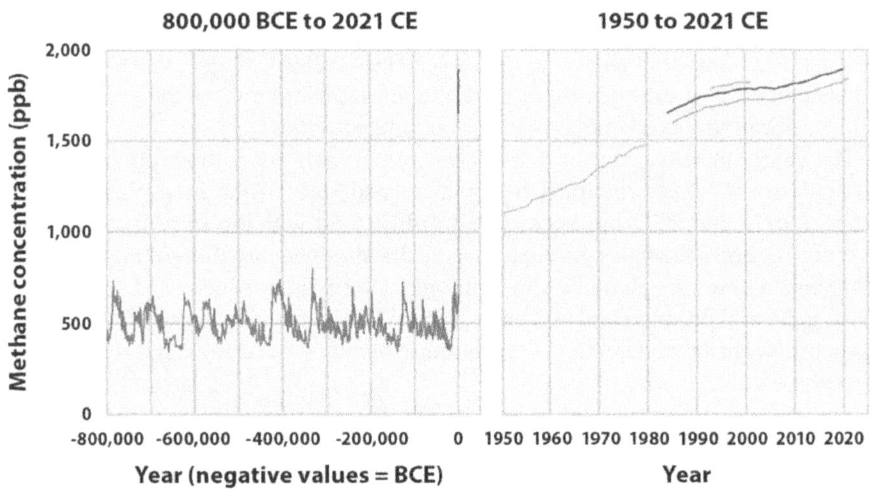

Fig. 5.5 **a** Yearly variation of concentrations (ppm) of carbon dioxide (CO_2) in the atmosphere [7], **b** Yearly variation of yearly concentration of methane (ppb, parts per billion) in the atmosphere [7], **c** Yearly variation of yearly concentration of nitrogen oxide (N_2O) (ppb, parts per billion) in the atmosphere. One ppb means one molecule of nitrogen oxide (N_2O) gas in every 1 billion molecules of air [7], **d** Yearly variation of yearly concentration of several halogenated gases (ppt, parts per trillion) in the atmosphere [7], **e** Yearly variation of yearly concentration of ozone (O_3) (Dobson unit) in the atmosphere [The Dobson unit (DU) is a unit of measurement of the amount of a trace gas in a vertical column through the Earth's atmosphere. The one Dobson unit is defined as the thickness in units of 10 μm) of ozone (O_3) larer in vertical direction] [7].

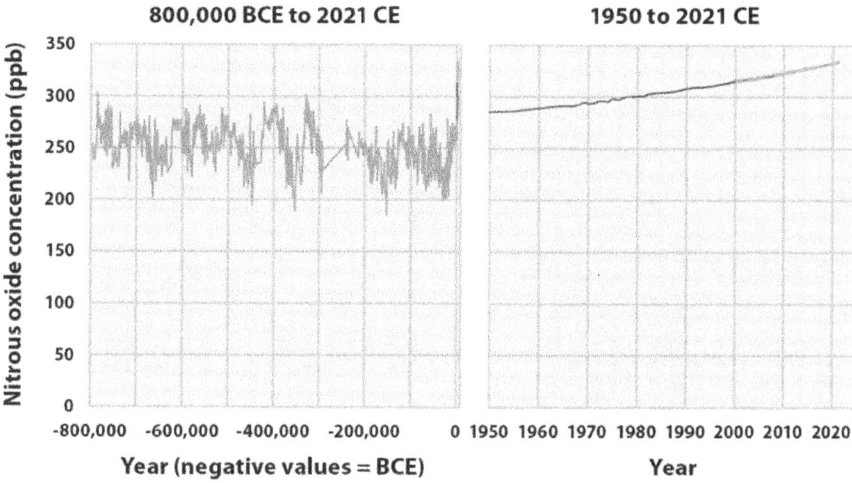

c

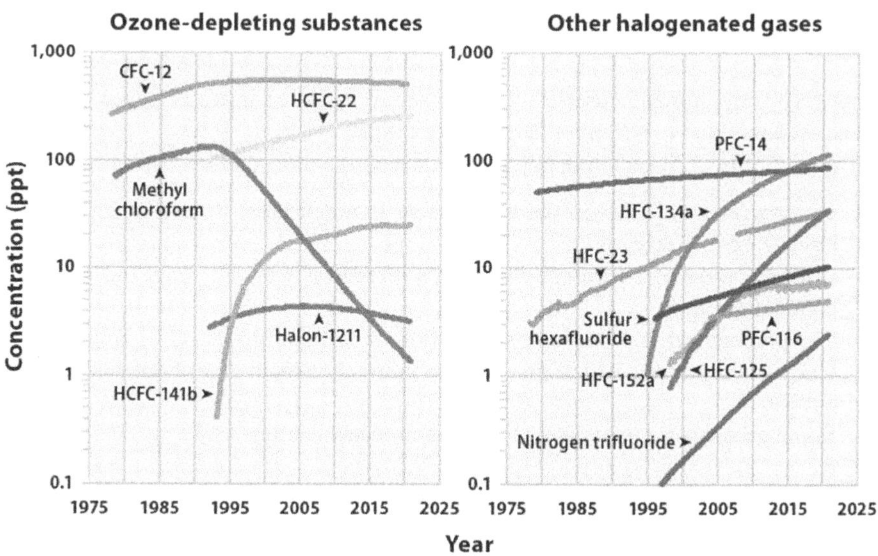

d

Fig. 5.5 (continued)

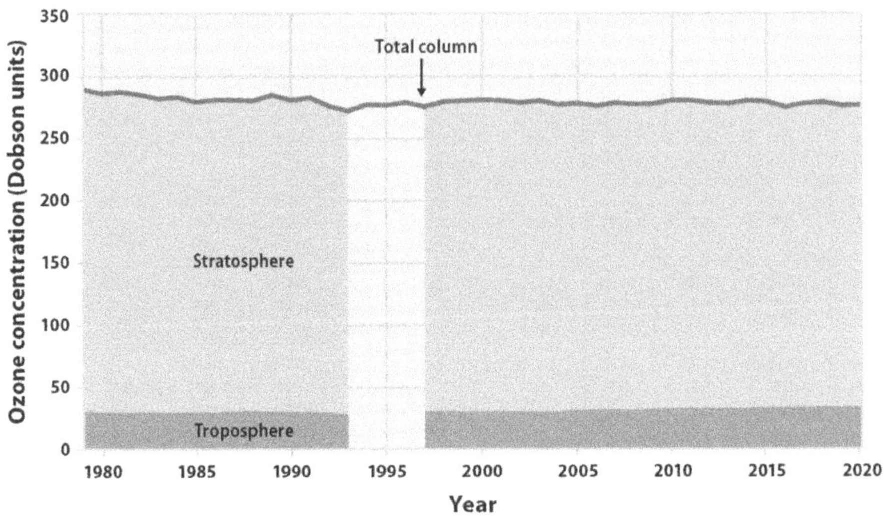

Fig. 5.5 (continued)

vapor (H_2O) by human activities. There is small variation in water vapor (H_2O) in the atmosphere due to irrigation and deforestation.

Based on Fig. 5.5a–e, some key points of the greenhouse effect are as follows:

(1) Concentration of CO_2 in the atmosphere is much more than concentration of other greenhouse gases.
(2) Concentration of all greenhouse gases increases over the years unlike ozone concentration.
(3) The percentage increase in concentration of carbon dioxide, CO_2 and methane CH_4 is about 50%; which much more than other main greenhouse gases such as NO_x.
(4) Some greenhouse were absent prior to industrialization.

5.4 Carbon Footprint (CF)

The carbon footprint (CF) is based on an amount of carbon dioxide (CO_2) emissions associated with all the activities of a human as shown in Fig. 5.6. It is generally expressed in tons of CO_2 or equivalent per year. It is broadly classified as follows:

(a) **Direct emission**: It is also referred as **scope-I**. It is from sources that are owned or controlled by the reporting entity. For example, the emission from direct use of fossil fuel combustion on site namely personal vehicles (oil)/gas burning stoves/cooking/and manufacturing any product heating, transportation, emission produced from power plant etc.

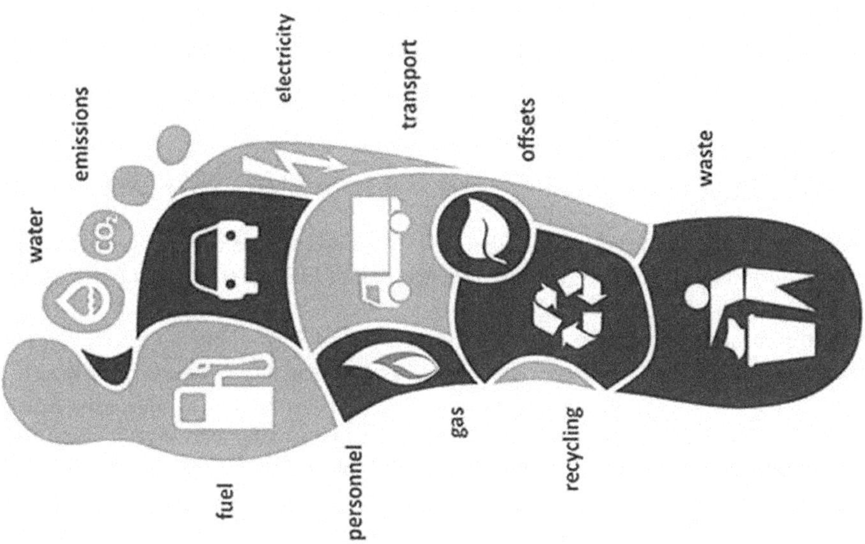

Fig. 5.6 Different components of carbon footprint

(b) **In direct emission**: It is also referred as **scope-II and scope-III**. It is a consequence of the activities of the reporting entity occurred at sources owned or controlled by another entity, Example in indirect emission are transportation of fossil/fuels, any other energy used outside of the production facility, wastes produced outside of the production facility, any end-of-life process or treatments (recycled processes), product and waste transportation and emissions associated with selling the product etc.

In addition to CO_2 emissions, it also includes the methane (CH_4), nitrous oxide (N_2O) and chlorofluorocarbons (CFCs) emissions. The Greenhouse Gas Emission Protocol (GHG-EP) has extended the range of seven gases by the Kyoto Protocol [8, 9] namely (i) carbon dioxide (CO_2), (ii) methane (CH_4), (iii) nitrous oxide (N_2O), (iv) hydrofluorocarbons (HFCs), (v) perfluorocarbons (PCFs), (vi) sulphur hexafluoride (SF6) and (vii) nitrogen trifluoride (NF3).

A carbon footprint is the measure of the total amount of greenhouse gas (GHG) emissions expressed as carbon dioxide equivalent/CO_2 equivalent/CO_2-eq (million metric tons of carbon dioxide equivalents, MMTCDE). It is caused by many factors like an individual, event organization, the burning of fossil fuels, land clearance and the production and consumption of food, manufacture of goods, materials, wood, roads, buildings, transportation and other services.

According to the Carbon Dioxide Information Analysis Center and the United Nations Development Programme, in 2004, the average resident had a per capita carbon footprint of the United States, France, Brazil and Tanzania of 20.6 metric tons (22.7 short tons), 6.0 metric tons (6.6 short tons), 1.8 metric tons (about 2 short tons) and 0.1 metric ton (0.1 short ton). (22.7 short tons) of CO_2 equivalent. The United States have five to seven times the global world, on average.

Carbon Footprint | Definition, Examples, Calculation, Effects https://www.britannica.com

The million metric tons of carbon dioxide equivalents (MMTCDE) can be evaluated from the following expression:

$$\textbf{MMTCDE} = (\text{million metric tons of a gas}) \times (\text{GWP of the gas}) \qquad (5.1)$$

where the global warming potential (GWP) depends (i) the absorption of infrared solar radiation by a given gas (ii) the time horizon of interest (integration period) and the atmospheric lifetime of the gas.

A high global warming potential (GWP), S.No. 4-7 of Table 5.1 correlates with a large infrared solar radiation absorption and a long atmospheric lifetime. It is important to note that (i) carbon dioxide (CO_2), (ii) methane (CH_4), (iii) nitrous oxide (N_2O) are the main greenhouse gases emitted during burning of fossil fuels for power generation, which is responsible for the deteriorating environment and climate change, Fig. 2.2b.

The methane (CH_4) is also converted into carbon dioxide (CO_2) as per reaction

$$CH_4 + 2O_2 = CO_2 + 2H_2O + \text{energy}(890\ \text{kJ/mol}) \qquad (5.2)$$

Example 5.1 Calculate million metric tons of carbon dioxide equivalents (MMTCDE) for methane (CH_4) and nitrogen oxide (N_2O).

Solution:

From Table 5.1, GWP for methane (CH_4) and nitrogen oxide (N_2O) are as follows:
GWP of methane (CH_4) = 25 and
GWP of nitrogen oxide (N_2O) = 298
By using Eq. 5.1, one gets
MMTCDE of CH_4 = (million metric tons of a gas) × 25 = 25 million metric tons of carbon dioxide and

Table 5.1 Global warming potential (GWP) for greenhouse emission [7]

S.No	Greenhouse gas	Chemical formula	100 year global warming potential
1	Carbon dioxide	CO_2	1
2	Methane	CH_4	24
3	Nitrous oxide	N_2O	298
4	Hydrofluorocarbons (HFCs), HFC-23	CHF_3	14,800
5	Perfluorocarbons, Carbon tetrafluoride—PFC-14	CF_4	7390
6	Sulphur hexafluoride	SF_6	22,800
7	Nitrogen trifluoride	NF_3	17,200

MMTCDE of N_2O = (million metric tons of a gas) × 298 = 298 million metric tons of carbon dioxide.

5.4.1 Carbon Footprint Calculation

There are many methods for calculating carbon footprint at various levels, such as household, company, national and international. We will discuss various option used in the literature, which are as follows:

5.4.1.1 Individual Household Carbon Footprint

In this case, yearly consumption of electricity, patrol, diesel and liquefied petroleum gas (LPG) should be known in individual family. The expression for each consumption are given below:

(i) **For electricity**: Output value in Kg of CO_2 = Input value in KWh/Yr × 0.85 (Emission Factor)
(ii) **For petrol**: Output value in Kg of CO_2 = Input Value in Liters/Yr × 2.296 (Emission Factor)
(iii) **For diesel**: Output value in Kg of CO_2 = Input Value in Liters/Yr × 2.653 (Emission Factor)
(iv) **LPG**: Output value in Kg of CO_2 = Input Value in Kg/Yr × 2.983 (Emission Factor).

$$\textbf{Total yearly carbon Footprint in tons} = \frac{\text{Sum of}(1+2+3+4)}{1000} \tag{5.3}$$

Example 5.2 Evaluate carbon foot print for a family for the following parameters:

(a) Yearly electricity consumed = 3000 kWh
(b) Yearly petrol consumptions = 6000 L
(c) Yearly diesel consumptions = 9000 L
(d) Yearly LPG consumptions = 360 kg

Solution:
By using Eq. 5.2 one has

$$\text{Yarelycarbon footprint in tons} = \frac{\begin{array}{c}3000 \times 0.85 + 6000 \times 2.296 \\ +9000 \times 2.653 + 360 \times 2.983\end{array}}{1000}$$
$$= \frac{2940 + 13,776 + 23,877 + 1,074}{1000}$$

$$= 41.667$$

Ref: Carbon Footprint Calculator—Calculate Your CO_2 Emissions https://co2.myclimate.org/calculate/emissions.

5.4.1.2 Alexandra Shimo-Barry Formula

He has developed another formula to compute carbon footprint and it is given by the following steps:

(j) Multiply your yearly electric bill by 105.
(ii) Multiply your yearly gas bill by 105.
(iii) Multiply your yearly oil bill by 113.
(iv) Multiply your total yearly mileage on your car by 0.79.
(v) Multiply the number of flights you've taken in the past year (4 h or less) by 1,100.
(vi) Multiply the number of flights you've taken in the past year (4 h or more) by 4,400.
(vii) Add 184 if you do not recycle newspaper.
(viii) Add 166 if you do not recycle aluminum and tin.
(ix) Add (i–viii) together for your total carbon footprint.

The following are the range of carbon footprint:

1. Low carbon footprint lies in the range of between 6,000 and 15,999 pounds.
2. Average carbon footprint lies in the range of 16,000–22,000 pounds.
3. Very low carbon footprint lies below 6000 pounds.
4. With living green practices means forestation carbon foot print lies over 22,000 pounds.

Example 5.3 Evaluate carbon footprint by using Alexandra Shimo-Barry formula for the following parameters:

(a) Yearly electricity bill in rupees = 18,000/-
(b) Yearly petrol bill in rupees = 18,000
(c) Yearly diesel consumptions = 60,000
(d) Yearly LPG in kg = 12,000

Solution:
By using Alexandra Shimo-Barry formula, one has following carbon footprint

$$\begin{aligned}\text{Yearly carbon footprint in rupees} &= 18,000 \times 105 + 6,000 \times 113 \\ &\quad + 9,000 \times 113 + 12,000 \times 105 \\ &= 1,854,000 + 1,695,000 + 1,260,000 \\ &= \text{Rupees } 4,809,000 = 48,090 \text{ pound}\end{aligned}$$

5.4.1.3 Method for Evaluating Carbon Footprint

Calculation of carbon footprint based on [References: Carbon footprint calculation formulas—IBM, https://www.ibm.com and CO_2 Emissions from Stationary Combustion of fossil fuel, [https://www.ipcc-nggip.iges.or.jp/bgp].

Figure 5.6 shows the various parameters, namely travel (car, plane, train, etc.), electrical use, consumption of products and services, foods, etc. involved to calculate carbon footprint. Before calculating, one should know some factors which will be used for analyzing the carbon footprint, which are as follows:

(a) **Oxidization Factor (F_{ox})**
During combustion of fossil fuel, all carbon is not oxidized due to deposition of carbon in the equipment of burning the fossil fuel, its exhaust flues, or carried within its gases or remain in residue and ashes. The fraction of fossil fuels which is oxidized during combustion is known as oxidation factor (F_{ox}). For some fossil fuels, the oxidization factor (F_{ox}). is as follows:

(i) **Coal**: 0.98
(ii) **Oil products (petrol and diesel)**: 0.99
(iii) **Natural gas**: 0.995.
Since there is little variation, this is always considered as one but it should be taken into account.

(b) **Carbon storage factor (F_{st})**
The carbon storage factor (F_{st}) is a fraction and it is ratio of unutilized carbon to total burned fuel. For complete combustion/usage, it is zero. It is used to discount part of the carbon content of the fuel. The part discounted is deemed to be equal to the carbon stored within the long-life products manufactured from the fuel. For example, a carbon storage factor for lubricants is to reflect the carbon remaining in waste oils and any recycled quantities.
The IPCC provide storage factors (F_{st}) for coal, diesel oil, and Natural gas and LPG as follows:

(a) **For coal, $F_{st} = 0$**
(b) **For coal tar, $F_{st} = 1$**
(c) **For oil, $F_{st} = 0.5$**
(d) **For natural gas, $F_{st} = 0.33$**

There are many expressions proposed by various researchers on carbon footprint, however, we are presenting in simplified expression for carbon footprint. These are as follows:

(a) **Carbon footprint calculations for energy**

Scope-I

The following calculations are used for direct energy emissions:

$$\textbf{CO}_2\textbf{ in tons} = [(\text{Total amount in natural unit (kg)} \times \text{Heating Value in kWh per natural unit}) \{\text{CO}_2 \text{ Emission factor} \times (1 - F_{st}) \times F_{ox} \times \text{GWP CO}_2 \text{ conversion}]/1000 \quad (5.4a)$$

$$\textbf{CH}_4\textbf{ in tons} = [(\text{Total amount in natural unit (kg)} \times \text{Heating Value in kWh per natural unit}) \{\text{CH}_4 \text{ Emission factor} \times (1 - F_{st}) \times F_{ox} \times \text{GWP CH}_4 \text{ conversion}]/1000 \quad (5.4b)$$

$$\textbf{N}_2\textbf{O in tons} = [(\text{Total amount in natural unit (kg)} \times \text{Heating Value in kWh per natural unit}) \{\text{N}_2\text{O Emission factor} \times (1 - F_{st}) \times F_{ox} \times \text{GWP N}_2\text{O conversion}]/1000 \quad (5.4c)$$

where GWP for greenhouse gases and heating value, emission factor for different fuel are given in Table 5.2 respectively. The values of F_{ox} and F_{st} are given above.

Table 5.2 Heating value and emission factor of different fuel [(a) 1 mmbtu = 295 kWh; (b) 1 kg CO_2 per mmBtu = 1/295 kg CO_2/kWh = 0.00339 kg/kWh; (c) 1 mmBtu per gallon = 295 kWh/3.79kg = 77.83 kWh/kg]

S.No	Fuel type	Heating value (kWh/kg)	CO_2 factor (kg/kWh)	CH_4 factor $\times 10^{-3}$ (kg/kWh)	N_2O factor $\times 10^{-3}$ (kg/kWh)
1	Hard coal	8.06	0.383	0.0373	0.0054
2	Petroleum (liquid)	11.13	0.347	0.0102	0.002034
3	Natural gas	11.76	0.17967	0.00339	0.000339
4	Vegetable oil		0.2765	0.00373	0.000373
5	Lignite coal	4.61	0.331	0.0373	0.0054
6	Wood and wood residuals	5.68	0.318	0.0244	0.0122
7	Crude oil	11.12	0.253	0.0102	0.00203
8	Kerosene oil	10.51	0.1939	0.0102	0.00203
9	Used oil	10.74	0.2508	0.0102	0.00203
10	Biodiesel (100%)	9.96	0.2503	0.00373	0.0054
11	Ethanol	6.54	0.232	0.00373	0.0054
12	Biogas	5.56	0.251		

Scope-II

The following calculations are used for indirect energy emissions:

$$\textbf{CO}_2\textbf{ in tons} = [\text{Total amount used based on occupancy in kWh} \times \text{CO}_2\text{ Emission factor} \times \text{GWP CO}_2\text{ Conversion}]/1000 \quad (5.5a)$$

$$\textbf{CH}_4\textbf{ in tons} = [\text{Total amount used based on occupancy in kWh} \times \text{CH}_4\text{ Emission factor} \times \text{GWP CH}_4\text{ Conversion}]/1000 \quad (5.5b)$$

$$\textbf{N}_2\textbf{O in tons} = [\text{Total amount used based on occupancy in kWh} \times \text{N}_2\text{O Emission factor} \times \text{GWP N}_2\text{O Conversion}]/1000 \quad (5.5c)$$

(b) **Carbon footprint calculations for travel**

Scope-I

The following calculations are used for direct travel emissions for company vehicles

$$\textbf{CO}_2\textbf{ in tons} = [\text{Amount of fuel in kg} \times \text{Heating Value} \times \text{CO}_2\text{ Emission factor} \times (1 - F_{st}) \times F_{ox} \times \text{GWP CO}_2\text{ conversion}]/1000 \quad (5.6a)$$

$$\textbf{CH}_4\textbf{ in tons} = [\text{Amount of fuel in kg} \times \text{Heating Value} \times \text{CH}_4\text{ Emission factor} \times (1 - F_{st}) \times F_{ox} \times \text{GWP CH}_4\text{ conversion}]/1000 \quad (5.6b)$$

$$\textbf{N}_2\textbf{O in tons} = [\text{Amount of fuel in kg} \times \text{Heating Value} \times \text{N}_2\text{O Emission factor } (1 - F_{st}) \times F_{ox} \times \text{N}_2\text{O Emission factor}]/1000 \quad (5.6c)$$

Scope-II

The following calculations are used for direct travel emissions for own vehicles

$$\textbf{CO}_2\textbf{ in tons} = [\text{Fuel Usage in kWh} \times \text{CO}_2\text{ Emission factor} \times F_{ox} \times \text{GWP CO}_2\text{ Conversion}]/1000 \quad (5.7a)$$

$$\textbf{CH}_4\textbf{ in tons} = [\{\text{Fuel Usage in kWh} \times \text{CH}_4\text{ Emission Factor}\} \times F_{ox} \times \text{GWP CH}_4\text{Conversion}] \quad (5.7b)$$

$$\textbf{N}_2\textbf{O in tons} = [\{\text{Fuel Usage in kWh} \times \text{N}_2\text{O Emission Factor}\}$$

$$\times F_{ox} \times \text{GWP N}_2\text{O Conversion})] \tag{5.7c}$$

From Table 5.2, it is to be noted that emission factor for methane (CH_4) and nitric oxide (N_2O) is very small in comparison to emission factor for carbon oxide (CO_2) and hence it can be neglected for calculating carbon footprint. Further for coal, oxidization factor, F_{ox} and GWP CO_2 conversion is approximately one and storage capacity is zero hence finally carbon footprint expression becomes

$$\begin{aligned}\textbf{CO}_2\textbf{ in tons} = & [(\text{Total amount in natural unit (kg)} \\ & \times \text{Heating Value in kWh per natural unit, kg)} \\ & \times \text{CO}_2\text{ Emission factor}]/1000\end{aligned} \tag{5.8}$$

The above equation is similar to those used in Example 5.2. In Example 5.2, emission factor for various fuel is more than reported value in Table 5.2 due to open burning of fossil fuel for domestic purposes.

If one desire to evaluate the quantity of carbon (C) in tons released to atmosphere due to combustion of fossil fuel, the amount of CO_2 in tons should be obtained as

$$\text{Carbon (C) in tons} = \text{Amount of CO}_2\text{ in tons} \times \frac{12}{44} \tag{5.9}$$

Example 5.4 Evaluate carbon footprint by using Eq. 5.4a and carbon content in CO_2 for input coal of amount 2500 metric tons for power generation.

Solution:

Amount of coal = 2500 metric tons = 2500 × 1000 kg
Heating value of coal = 8.06 kWh/kg (Table 5.2)
CO_2 emission factor = 0.383 kWh/kg (Table 5.2)

$$F_{st} = 0.10\text{, assumed and } F_{ox} = 0.98$$

CO_2 GWP conversion = 1 (Table 5.1)
Substitute above numeric value in Eq. 5.4a, we get

$$\begin{aligned}\textbf{CO}_2\textbf{ in tons} & = [(2,500 \times 1,000 \times 8.06) \times 0.383 \times (1 - 0.10) \times 0.98 \times 1]/1,000 \\ & = 6,806.79\text{ tons}\end{aligned}$$

Carbon content in CO_2 = 6,806.79 × $\frac{12}{44}$ = 1,856.40 tons (Eq. 5.9).

Example 5.5 Evaluate carbon footprint by using Eq. 5.8 and carbon content in CO_2 for input coal of amount 2500 metric tons for power generation.

Solution:

Amount of coal = 2,500 metric tons = 2,500 × 1,000 kg (Table 5.3)
Heating value of coal = 8.06 kWh/kg (Table 5.2)
CO_2 emission factor = 0.383 kWh/kg (Table 5.2)

Table 5.3 Densities of greenhouse gases (GHG) and some fuels

S.No	Greenhouse gas	Chemical formula	Molecular weight, gm/mole	Density, ρ (kg/m^3)
1	Carbon dioxide	CO_2	44.01	1.9095
2	Methane	CH_4	16.043	0.6925
3	Nitrogen oxide	N_2O	44.013	1.34
4	Ozone	O_3	48.0	2.14
5	Water vapor	H_2O	18.016	0.804
6	Sulfur dioxide	SO_2	64.06	2.6025
7	Oxygen	O_2	32	1.38
8	Hydrogen sulfide	H_2O	34.076	1.434[1]
9	Carbon monoxide	CO	28.01	1.205
10	Petrol	C_nH_{2n+2}	220–238	730–770
11	Diesel	$C_{10}H_{20}$	198–202	830
12	Coal	$C_{137}H_{97}O_9NS$	3142.32	1000
13	Natural gas	CH_4	19.0	0.68
14	Kerosene	$C_{12}H_{26}$	170	775–840

Substituting the above numeric values in Eq. 5.4a, we get

CO_2 in tons = [(2,500 × 1,000 × 8.06) × {0.383]/1,000 = 7,717.45 tons

So, there is an increase of 11.8% in evaluating of carbon footprint between Examples 5.4 and 5.5.

Carbon content in $CO_2 = 6{,}806.79 \times \frac{12}{44} = 2104.76$ tons (Eq. 5.9).

Similarly, there is an increase of 11.8% in evaluating of carbon content in CO_2 between Examples 5.4 and 5.5.

Example 5.6 Evaluate the amount of fossil fuel for a given carbon footprint of 6,806.79 tons by using Eq. 5.4a for power generation

Solution:

Heating value of coal = 8.06 kWh/kg (Table 5.2).

CO_2 emission factor = 0.383 kWh/kg (Table 5.2).

$$F_{st} = 0.10, \text{ assumed and } F_{ox} = 0.98$$

CO_2 GWP conversion = 1 (Table 5.1).

Substitute above numeric value in Eq. 5.4a, we get

$$\textbf{CO}_2\textbf{ in tons} = \text{Amount of fossil fuel} \times 8.060.383 \times (1 - 0.10) \times 0.98 \times 1]/1,000$$
$$= 6,806.79 \text{ tons}$$

$$\text{Amount of fossil fuel} = \frac{\mathbf{CO_2 \text{ in tons}} \times 1,000}{8.06[0.383 \times (1 - 0.10) \times 0.98 \times 1]}$$
$$= 2,499,999 \text{ tons} = 2.499 \text{ metric tons} \approx 2.5 \text{ mt}$$

The densities of some greenhouse gases (GHG) and fuels are given in Table 5.3.

5.5 Net-Zero Emissions

The Paris Agreement (PA) underlines the need for net-zero greenhouse gas (GHG) emission. Net-zero greenhouse gas (GHG) emission means a condition in which the greenhouse gases, carbon dioxide (CO_2), methane (CH_4) and nitrous oxide (N_2O), are balanced by removal of these gases from the atmosphere. Human activities include deforestation, land use and land-use changes (LULUC), livestock production, fertilization, waste management and industrial processes.

5.5.1 Net-Zero Energy Building

The Net-Zero-Energy Building (N-ZEB) concept is now a days gaining momentum across the globe to meet the energy savings with net-zero emission of carbon dioxide (CO_2). It is important for us to know the need of net-zero energy buildings (N-ZEB). This is because humans live on the planet Earth, which is the only planet that is inhabitable. Human are constantly disturbing the Earth to meet their energy demands and today, humans have to face the following sustainability climate issues. The first issue is the climatic changes which are already happening and we need to be prepared for extreme weather conditions and become climate adaptive. The aggravated climate change is caused by greenhouse gases (GHG), mostly emitted through human processes as mentioned earlier. Estimates are varying, but the temperature on earth will most probably rise between 1.5 and 5 degrees within this century, depending on the action taken or not [10]. The second reason is the depletion of fossil fuels, namely coal, petroleum and natural gas. In order to avoid these climatic changes, we need to shift to renewable sources of energy (Chap. 3). The third reason is the vulnerability. This means that the urbanized areas have been experiencing disturbances in the supply of energy over a last few decades. Therefore, in order to create a resilient society, our cities have to become more self-sufficient in terms of energy at least.

The only way to solve this is to become entirely circular in our material use. Carbon emissions must be reduced, necessitating a rapid energy transition to clean renewable energy.

The built environment constitutes an important factor because it involves the use of 30–40% of all energy, thus there is a need for it to become carbon neutral.

To achieve this, there is a need to design zero- energy buildings and also to make existing structures energy neutral [11].

5.5.2 The Zero-Energy Building (ZEB) [12]

There is a special relationship between the design of buildings and natural processes that offer the potential for inexhaustible sources of energy. The concept is not new and various examples from the vernacular architecture can be traced. The definition of zero energy building is to be flexible and able to assess the quality of energy used for heating/cooling of buildings, different climatic conditions, types of buildings and existing building masses.

The Zero-Energy Building (ZEB) has different definitions associated with it. One can talk of energy neutral, zero energy, or net zero energy. In the Netherlands, the term zero on the meter is also well-known, indicating that over a year's time there are no energy costs.

The zero energy building may be defined as the one which does not consume any fossil fuels while its construction. In Indian history, the architecture has been evolved with time, depending upon the traditional building materials, technologies and craftsmanship available. A kachcha building (mud house) was referred to the construction done using building materials like mud, grass, bamboo, thatch or sticks as shown in Fig. 5.7. In this case, there is zero embodied energy involved in materials, namely mud, grass, bamboo, thatch or sticks used for construction of mud houses (see Fig. 5.7a). A mud brick is a fire-free brick, made of a mixture of clay, mud, sand, and water mixed with a binding material such as rice husks or straw. They use a stiff mixture and let them dry in the sun for 25 days. Adobe is a type of mud brick also used today to save energy and is an environmentally safe way to insulate a house (Fig. 5.7a(iv)). In addition its has a cooling effect during summer period which is needed for thermal comfort.

Next are brick buildings. There are additional materials such as cow dung, mud plaster etc. required for construction (see Fig. 5.7a). In this case, embodied energy is not zero due to the use of bricks in such a house, but energy payback time (EPBT) is small; even less than one year. These raw materials were weak; durable materials are used in kachcha house (mud house) construction. In this building, none of any external sources of energy are needed to provide thermal comfort inside building. The brick buildings are a step towards Net Zero Energy Building (N-ZEB) (see Fig. 5.7b). Embodied energy (Sect. 4.5.1) in ZEB is zero and when this embodied energy tends to increase, and ZEB moves towards the Net Zero Energy Building (N-ZEB) concept. Embodied energy may be defined as quantity of energy required by all activities; i.e., the overall expenditure of energy required to construct and maintain the building over the whole life cycle.

The concept of zero energy building (ZEB) has been definedby various researchers [11] on the basis on energy use.

Fig. 5.7 **a** (i) Photograph of grass and bamboo house (ii) mud house (iii) A photograph of a kachcha building (mud house) and (iv) Mud brick, **b** A photograph of brick building with day light arrangement through roof

(i)

Semi Urban Housing (Jaipur)

(ii)

(iii)

(iv)

a

b

Fig. 5.7 (continued)

- As suggested by Buildings Performance Institute Europe (BPIE), "nearly zero energy building should include a threshold for household electricity used for integrated building equipment (like lifts etc.) going beyond the building services (heating, cooling, ventilation and lighting)" [13].
- According to Gilijamse [14] a ZEB as "a house in which no fossil fuels are consumed, and the annual electricity consumption equals annual electricity production. Unlike the autarkic situation, the electricity grid acts as a virtual buffer with annually balanced deliveries and returns".
- It has also been defined by Torcillini et al. [15] as "a residential or commercial building with greatly reduced energy needs through efficiency gains such that the balance of energy needs can be supplied with renewable technologies".
- With reference to the above definition made by Torcillini et al. [15], Kilkis [16] came up with the concept of Net-Zero Energy building and defined it as "a building which has a total annual sum of zero energy transfer across the building-district boundary in a district energy system, during all electric and any other transfer that is taking place in a certain period of time".
- Another definition for ZEB was given in the report by International Energy Agency, a report by Laustsen [17] which states "Zero energy buildings do not use fossil fuels but only get all their required energy from solar energy and other renewable sources". Marszal and Heiselberg [18] have very well indicated the key pointers to define a ZEB.
- According to [19] ZEB consists of two strategies—minimizing demand through energy-efficient measures (passive design) and using renewable energy and other technologies to meet remaining energy requirements (active design). First is the passive strategies which can reduce the energy demand of the ZEB as much as possible. The passive design strategies including shading, now-thermal bridge, air tightness design, natural ventilation and lighting, and high-performance envelope (see Fig. 5.8). In this case, thick white marble (Trombe wall concept), increased height, dome shape, underground structure etc. have been used. Due to the white marble and dome shape, most of incident solar radiation is reflected to minimize heat to be absorbed. The other one is using high-efficient active technologies for renewable energy. Efficient lighting, energy efficient appliances, and fresh air

Fig. 5.8 Net-zero energy Taj Mahal, Agra, India

heat recovery system are included in the active design. The cost-optimal model of ZEBs based on passive and active technologies should be evaluated carefully. The technical guidelines stipulate the minimum requirements for ZEBs to meet the relevant energy consumption indicators, air tightness indicators and indoor environmental parameters.

Problems

Problem 5.1 Calculate methane (CH_4) in tons for Example 5.4 and 5.5 and compare it results with the carbon96 footprint obtained in the same Examples.

Hint: See Example 5.4 and use Eq. 5.4b and Tables 5.1, 5.2 and 5.3.

Problem 5.2 Calculate nitric oxide (N_2O) in tons, for Example 5.4 and 5.4, and compare results with carbon footprint obtained in the same Examples.

Hint: See Example 5.4 and use Eq. 5.4c and Tables 5.1, 5.2 and 5.3.

Problem 5.3 Calculate total carbon footprint of Examples 5.4 and 5.5 and Problems 5.1 and 5.2.

Hint: Add carbon footprint obtained in Examples 5.4 and 5.5 and Problems 5.1 and 5.2.

Problem 5.4 Calculate total carbon footprint for fossil fuels namely petroleum and natural gas for 10 and 20 m^3 for scope–I.

Hint: See Example 5.6.

Problem 5.5 Calculate total carbon footprint for fossil fuels namely petroleum and natural gas for 8,300 and 14 metric tons for scope–I.

Hint: Use Eq. 5.6 and Table 5.3.

Problem 5.6 Calculate total carbon footprint for fossil fuels namely petroleum and natural gas for 8,300 and 14 metric tons for scope–II.

Hint: Use Eq. 5.6 and Table 5.3.

Problem 5.7 Calculate total carbon footprint for fossil fuels petroleum and natural gas for 1000 L and 10,000 L.

Hint: Use Eq. 5.6 and Table 5.3.

Objective Questions

5.1 Global warming potential (GWP) for greenhouse emission is maximum for

(a) Coal
(b) Sulphur hexafluoride
(c) Nitrous oxide
(d) Methane

Answer: (b)

5.2 Global warming potential (GWP) for greenhouse emission is minimum for

(a) Coal
(b) Sulphure hexafluoride
(c) Nitrous oxide
(d) Methane

Answer: (a)

5.3 The global warming potential (GWP) for greenhouse emission for coal is

(a) One
(b) Hundreds
(c) 0.1
(d) 0.001

Answer: (a)

5.4 The maximum heating value among coal, petroleum, natural gas and Lignite coal is for

(a) Coal
(b) Petroleum

(c) Natural gas
(d) Lignite

Answer: (c)

5.5 Maximum density of fossil fuel is for

(a) Petroleum
(b) Natural gas
(c) Coal
(d) All of these

Answer: (c)

5.6 Oxidization factor (F_{ox}) is one for

(a) Coal
(b) Petroleum
(c) Natural gas
(d) All of these

Answer: (d)

5.7 Storage factors ($\boldsymbol{F_{st}}$) for coal is

(a) Ten
(b) Zero
(c) Five
(d) None of these

Answer: (b)

5.8 Storage factors ($\boldsymbol{F_{st}}$) for coal tar is

(a) Ten
(b) Zero
(c) Five
(d) One

Answer: (d)

5.9 The mud house as shown in Fig. 5.7i is known as

(a) Zero energy building
(b) Net-zero energy building
(c) Energy intensive building
(d) All of these

Answer: (a)

5.10 The grass house as shown in Fig. 5.7ii is known as

(a) Zero energy building
(b) Net-zero energy building

(c) Energy intensive building
(d) All of these

Answer: (a)

5.11 The house shown in Fig. 5.8 is known as

(a) Passive house
(b) Energy efficient house
(c) Active house
(d) Hybrid house

Answer: (a) and (b)

5.12 White painting on outside building is most suitable for

(a) Winter condition
(b) Rainy season
(c) Summer condition
(d) All of these

Answer: (d)

5.13 Black painting and glazed on outside building is most suitable for

(a) Winter condition
(b) Rainy season
(c) Summer condition
(d) All of these

Answer: (a)

5.14 The mmbtu stands for

(a) Metric million British thermal unit
(b) Metric million Bangladesh thermal unit
(c) Metric million Bangalore thermal unit
(d) All of these

Answer: (a)

5.15 One mmbtu is equal to

(a) 295 kWh
(b) 295 Wh
(c) 295 W
(d) None of these

Answer: (a)

5.16 One mega joule is equal to

(a) 28 kWh
(b) 2.8 kWh
(c) 0.028 kWh
(d) 0.28 kWh

Answer: (d)

References

1. Climate Change and Green Energy—Sutori. https://www.sutori.com
2. Climate change—Wikipedia. https://en.wikipedia.org
3. IPCC SR15 Ch2 2018, pp. 95–96
4. IPCC SRCCL Summary for Policymakers 2019, p. 18
5. Ocean acidification and coral bleaching, Earth Journalism Network. 09 June 2016. https://earthjournalism.net
6. Coral bleaching—Wikipedia. https://en.wikipedia.org
7. Climate change indicator: Atmospheric Concentrations of Greenhouse Gases. US EPA epa. https://www.epa.gov
8. Carbon footprint—Wikipedia. https://en.wikipedia.org
9. Carbon footprint of electricity generation. https://www.parliament.ukpostpostpn268
10. IPCC. AR5 Synthesis Report: Climate Change 2014 (2014)
11. A.v.d. Dobbelsteen, Introduction, in *Zero-Energy Design: An Approach to Make Your Building Sustainable*, EDX (2020)
12. G.N. Tiwari, Neha Gupta, *Photovoltaic Thermal Passive House System* (Taylor & Francis Group and CRC Press, 2023)
13. Ecofys, *Towards Nearly Zero-energy Buildings*. Am Wassermann 36: ECOFYS Germany (2013)
14. W. Gilijamse, *Zero-Energy Houses in the Netherlands* (Madison, Wisconsin, USA, 1995)
15. P. Torcellini, S. Pless, M. Deru, *Zero Energy Buildings: A Critical Look at the Definition* (National Renewable Energy Laboratory, USA, 2006)
16. S. Kilkis, *A New Metric for Net-Zero Carbon Buildings* (Long Beach, California, 2007)
17. J. Lautsen, *Energy Efficiency Requirements in Building Codes, Energy Efficiency Policies for New Buildings* (International Energy Agency (IEA), 2008)
18. A. Marszal, P. Heiselberg, A literature review of zero energy building (ZEB) definitions (2009)
19. Z. Liu, Q. Zhou, Z. Tian, B.-j. He, A comprehensive analysis on definitions, development, and policies of nearly zero energy buildings in China. Renew. Sustain. Energy Rev. **114** (2019)

Related Other References

20. Carbon footprint—Wikipedia. https://en.wikipedia.org
21. Carbon footprint: a study of unit and dimensions. http://dimacs.rutgers.edu/mpe/Carbon

Chapter 6
Ecology

6.1 Introduction

The history of life on the planet Earth is interaction between living things including humans and the environment (see Fig. 6.1a). Ecology is the study of all living organisms (part of biology) including wildlife humans, associated with the surrounding environment of the planet Earth.

The strong relationship between all living things including wildlife and human on the planet Earth and their surroundings environment is referred to as Ecology.

The study of ecology is science including human science, population, community, ecosystem and biosphere with interaction among them. The main goal of an ecologist is to improve the understanding of living organism of (a) their life processes, (b) adaptations and habitats and (c) interactions and biodiversity.

6.2 Classification of Ecology

There are many types of ecology. Some of them are explained briefly as follows.

(a) **Human ecology**: The study of human behavior and its origin and development, their effect on each other and other living organism including wildlife and the environment. Humans exeter an enormous impact on the globe and the biome (forest or tundra, such as the large flat Arctic regions). Human ecology has also been defined as a type of analysis applied to the plants and animals in ecology.

(b) **Population ecology**: A sub-field of ecology. The population ecology is a group of organisms of the same species. The change in populations depends on birth and death rates, and by immigration and emigration.

G. N. Tiwari, *Energy, Ecology and Environment*,
https://doi.org/10.1007/978-981-99-3997-8_6

a

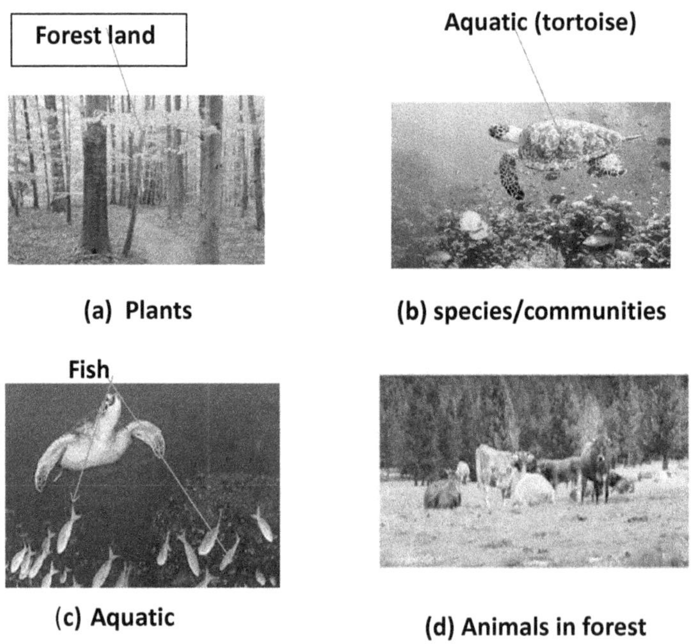

b

Fig. 6.1 **a** Theme of nature. **b** Few species on the planet Earth. **c** Classification of some of ecology. **d** Biosphere. **e** Five Indian tropical forest

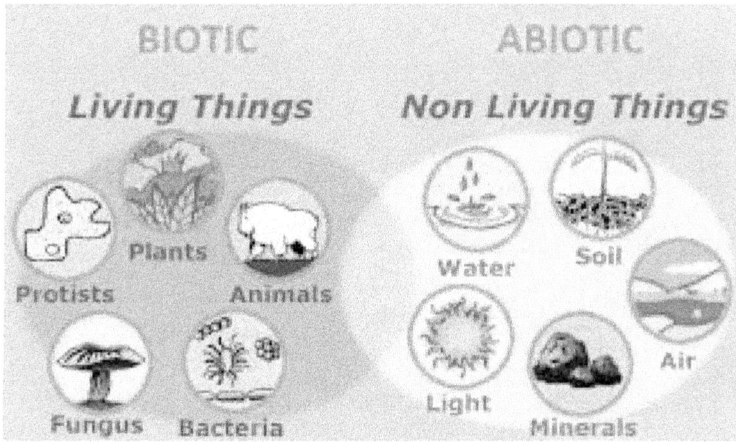

c

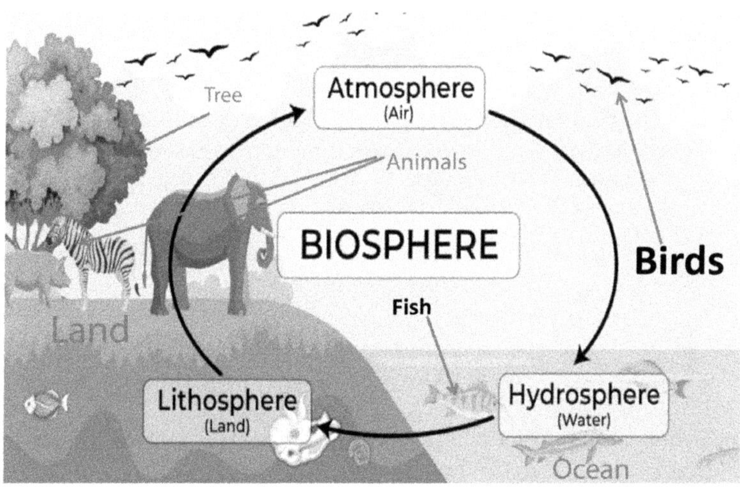

d

Fig. 6.1 (continued)

e

Fig. 6.1 (continued)

If $N(t)$ is the total number of individual species of living organisms on the planet Earth between 0 to t time interval then its expression at time 't' can be written as follows:

$$N(t) = N_0 + B(t) - D(t) + I(t) - E(t) \tag{6.1a}$$

or,

$$N(t) = N_0 + [B(t) + I(t)] - [D(t) + E(t)] \tag{6.1b}$$

where N_0 is the number of individual species at time 0, $B(t)$ = the number of individual species born between 0-t time interval, $D(t)$ = the number that died between 0-t time interval, $I(t)$ = the number of individual species immigrated between same period (The number of individuals that join a population over time), and $E(t)$ = the of individual species emigrated between time 0 and time t (The number of individuals that leave a population over time t).With reference to Eq. 6.1a, b, one can observe the following:

(i) It will be easy to have data for human and wildlife whose populations can be counted and collected regularly.
(ii) It is difficult to count and collect data for flying birds and aquatic animals.
(iii) However, it can be done for known sampling data.

It is important to mention that there should be conservation of populations covering all numbers of individual species for each organism; this shall be discussed later on.

Example 6.1a Derive/write an expression for birth/death rate along with population growth rate by using Eq. 6.1a, b.

Solution:

The birth rate (b) = $\frac{B(t)}{N_0}$; the death rate (d) = $\frac{D(t)}{N_0}$

Now the population growth rate per capita (r) can be evaluated as follows:

$$r_g = \frac{[B(t) - D(t)]}{N_0} = b - d \quad (6.1.1)$$

and an expression for the population decline rate per capita will be as follows:

$$r_d = -\frac{[D(t) - B(t)]}{N_0} = -(d - b) \quad (6.1.2)$$

Example 6.1b Determine the per capita growth rate and the actual number of the population of 1000 cow for annual growth and death of 15% and 10% respectively.

Solution::

Annual population growth rate per capita (r_g) = 0.15 – 0.10 = 0.05 [From Eq. 6.1.1].

The annual population growth of cow = 1000 × 0.05 = 50.

So, the total population of cow at end of year = 1000 + 50 = 1050.

Example 6.2a Derive an expression for exponential growth of a given species for Example 6.1a, b.

Solution:

For exponential growth, an expression can be written as

$$\frac{\Delta N}{N} = +r_g \Delta t$$

Take the integral on both side, it gives

$$\log \mathrm{N(t)} - \log N_0 == +r_g t \text{ or } \log \frac{N(t)}{N_o} = +r_g \mathrm{t} \text{ or } N(t) = N_0 e^{+r_g t} \quad (6.2.1)$$

where N_0 = initial population of given species at time t = 0 and **e = 2.71828** = Euler's number.

Similarly if population is in decline, the expression for decline population at time 't' becomes as follows:

$$N(t) = N_0 e^{-r_d t} \quad (6.2.2)$$

where an expression for r_g and r_d are given in Example 6.1a, b

The unit of 't' may be hourly, daily, monthly and yearly too depending upon the problem to be address.

Example 6.2b Determine population of a village at end of 5th for initial population (N_0) of 2500 and annual population growth (r_g) of 0.10

Solution:

With the help of Eq. 6.2.1, we have

Population of a village $= N(t) = N_0 e^{+r_g t} = 2500 \times e^{0.10\times 5} = 2500 \times 1.645 =$ 4,112

Example 6.3 Derive an expression for time period for population growth took place from N_0 to $N(t)$ for population growth of r_g by using 6.2a, b.

Solution:

From 6.2a, b, we have

$$N(t) = N_0 e^{+r_g t}$$

The above equation can be written as

$$\frac{N(t)}{N_0} = e^{+r_g t}$$

Take log of both side, one gets

$$\log\frac{N(t)}{N_o} = +r_g \mathrm{t} \quad \text{or} \quad \mathrm{t} = +\frac{1}{r_g}\log\frac{N(t)}{N_o} \tag{6.3.1}$$

Example 6.4 Find out an expression for population density for carrying capacity of 'C' for 6.2a, b

Solution:

In this case, an expression can be written as

$$\frac{\Delta N}{\Delta t} = +r_g\frac{C-N}{C} = r_g - \frac{r_g}{C}N \text{ or } \frac{\Delta N}{\Delta t} + \frac{r_g}{C}N = r_g$$

Multiply above equation by $e^{\frac{r_g}{C}t}$ on both sides, we have

$$e^{\frac{r_g}{C}t}\frac{\Delta N}{\Delta t} + e^{\frac{r_g}{C}t}\frac{r_g}{C}N = r_g e^{\frac{r_g}{C}t}$$

which can be rewritten as follows by replacing 'Δ' by 'd' deferential term:

$$e^{\frac{r_g}{C}t}\frac{dN}{dt} + e^{\frac{r_g}{C}t}r_g N = r_g e^{\frac{r_g}{C}t}$$

or,

$$\frac{dN}{dt}\left(Ne^{\frac{r_g}{C}t}\right) = r_g e^{\frac{r_g}{C}t}$$

After integration of above equations with initial conditions of $N = N_0$ at $t = 0$, We get

$$\left(Ne^{\frac{r_g}{C}t} - N_0\right) = C\left[e^{\frac{r_g}{C}t} - 1\right]$$

The solution becomes

$$\mathrm{N(t)} = \mathrm{C}\left(1 - \mathrm{e}^{-\frac{r_g}{C}t}\right) + \mathrm{N_0 e}^{-\frac{r_g}{C}t} \qquad (6.4.1)$$

Example 6.5 Discuss the limiting cases for Example 6.4
Solution:
From Example 6.4, we have

$$\boldsymbol{N(t) = C\left(1 - e^{-\frac{r_g}{C}t}\right) + N_0 e^{-\frac{r_g}{C}t}}$$

There are the following limiting cases:

Case (i): If $\boldsymbol{e^{-\frac{r_g}{C}t}} = 1$. This implies that either t or $r_g = 0$, for a given C. In this case $\mathrm{N(t)} = \mathrm{N_0}$: there is either no growth/decline or equal growth/decline in population of species
Case (ii): If $\boldsymbol{e^{-\frac{r_g}{C}t}}$tends zero (0). This implies that either t or r_g are very large (infinity) for given C. In this case, N(t) = C: there is no change in carrying capacity due to either very large population growth rate or large time interval.

(c) **Community population**: It is a group or association of populations of at least two or more different species existing/occupying/living together in a habitat (biotope) in the same geographical area at the same time. It is also known as a (i) biocoenosis, (ii) biotic (living) community, (iii) biological ecological community, (iv) Ecological community or life assemblage. In other words, it is groups of organisms in a specific place; for example, "the aquatic (fish) community of "a lake"."

Further, community ecology also takes into account of abiotic factors (non-living). It influences species distributions or interactions due to annual solar radiation, relative humidity, wind velocity air temperature or soil pH etc. For example, the living plant communities inhabiting (prohibit) deserts are very different in comparison with tropical rainforests due to differences in annual precipitation/rain fall. Humans can also affect community structure through habitat disturbance.

Each species occupies a niche (the role an organism plays) within the community for interaction with surrounding environment like day temperature and other climatic parameters and its role within the community. The different niches species can coexist. This is referred as niche partitioning which reduces competition between species. Due to coexistence, they suppress their own growth more than they limit the growth of other species. The competition within a species is greater than the competition between species. Some species has been shown in Fig. 6.1b. The number of niches present in a community determines the number of species present. If two species have the same food demands in the same niche then one species outcompetes (is more successful) than the other. The more niches filled, the higher the biodiversity (a reservoir of resources) of the community.

Example 6.5 Write down an expression for (a) population standard deviation and (b) sample standard deviation for growth of single species
Solution:
An expression for standard deviation can be evaluated from the following expressions:

(a) **Population standard deviation** (σ_{PSD})

$$\sigma_{PSD}(x) = \sqrt{\frac{\sum(x - \mu_p)^2}{N}} \tag{6.5.1}$$

where

$\sum$ is the summation of $\sum(x - \mu_p)^2$ from 1 to N (total population of given species)

x is individual numerical value of species at Nth number

$\mu_p = \frac{\sum x}{N}$ is the average value of total population

N is the total population of species.

(b) Sample Standard Deviation formula(σ_{SSD})

$$\sigma_{SSD}(x) = \sqrt{\frac{\sum(x - \mu_s)^2}{n - 1}} \tag{6.5.2}$$

where

$\sum$ is the summation of $\sum(x - \mu_s)^2$ from 1 to n (total number of samples of given species out of total population (N): n is less than N

$\mu_s = \frac{\sum x}{n}$ is the average value of sample population

n is the number of samples taken from total number of population (N).

Example 6.6 Write down an expression for co-variance and correlation of population and sample data's
Solution:
Covariance and correlation of two random species x and y is given by

(A) **Formula for Covariance**: It gives the direction of the relationship between two species

(a) For population of two species:

$$[Cov(x, y)]_{PCO} = \frac{\sum(x - \overline{x})(y - \overline{y})}{N} \tag{6.5.3}$$

where x and y are data for two species with average value of $\overline{x}$ and $\overline{y}$ for N population of each species.

(b) For sample for two from population data

$$[Cov(x, y)]_{SCO} = \frac{\sum(x - \overline{x})(y - \overline{y})}{n - 1} \tag{6.5.4}$$

where x and y are data for two species with average value of $\overline{x}$ and $\overline{y}$ from n sample taken out of N population. of each species.

(A) **Formula for correlation**: It measures the strength of the relationship between variables

(a) **For population of two species**:

$$[\rho(x, y)]_{PCOR} = \frac{[Cov(x, y)]_{PCO}}{\sigma_{PSD}(x)\sigma_{PSD}(y)}. \tag{6.5.5}$$

(b) **For sample from population data's of two species**:

$$[\rho(x, y)]_{SCOR} = \frac{[Cov(x, y)]_{SCO}}{\sigma_{SSD}(x)\sigma_{SSD}(y)} \tag{6.5.6}$$

(d) **Biodiversity**: Also referred as biological diversity. It is variety and variability of natural life and habitats on the planet Earth. It also measures the variation of (i) the genetic (genetic variability), (ii) species (species diversity), and (iii) ecosystem (ecosystem diversity) level. It is not equally distributed on the planet Earth. It is usually larger in the tropics with high primary productivity in the region near the equator due to the warm climate condition. Tropical forest ecosystems near the equator cover less than ten percent (10%) of earth's surface and have about ninety percent (90%) of the world's species. Similarly, marine biodiversity is usually higher along coasts in the Western Pacific (WP) near equator with high sea surface temperature (average annual temperatures >28 °C) and in the mid-latitudinal band in all oceans (The average temperature of the warmest month >10 °C and the coldest month < 30 °C). It encompasses the evolutionary, ecological, and cultural processes to sustain life.

(e) **Ecosystem**: It is a natural environment which includes plants (flora) and animals (fauna). A natural clean environment in biodiversity is essential to support all living organisms including human life. The natural environment becomes less productive, resilient and adaptable due to decrease in biodiversity and degradation of ecosystems. The environment is at real risk of sustaining long term damage or collapse under the circumstances mentioned above.

Nature plays an important role in climate change (Chap. 5) due to its degradation by human activity to meet demand like energy production, manufacturing and transport. All these human activities release carbon dioxide (CO_2) into the atmosphere. There are various natural processes for absorption of carbon dioxide (CO_2) from the atmosphere by plants and oceans. The degradation of natural habitats will disturb these processes and then accelerate to damage the climate, degradation of soil and underground levels of water etc.

Ecosystems are composed of a variety of biotic (living) and abiotic (non-living) components on the planet Earth. These function in an interrelated way (see Fig. 6.1c).

(f) **Biosphere**: The biosphere is also known as the ecosphere. Globally, is the sum of all ecosystems (plants (flora) and animals (fauna)). It is also termed as zone of life on the planet Earth. The biosphere consists of the parts of Earth with living organisms (where life exists) of all ecosystems which extends from the deepest root of plant (trees) to the dark environments of ocean trenches, to rainforests, high mountaintops, and transition zones where ocean and terrestrial ecosystems meet. The biosphere is a spherical shell like Earth. It is virtually a closed system, with minimal inputs and outputs as shown in Fig. 6.1d, and is an open system with respect to solar energy for photosynthesis which captures solar energy at a rate of around 1,000–1,100 W/m^2 in the terrestrial region (Chap. 1). The biosphere is the global ecological system consists of all living organisms and their relationships which includes their interaction with the the lithosphere, cryosphere (frozen water on the Earth), hydrosphere, and atmosphere, Fig. 6.1d. The biosphere is assumed to have evolved to create life (algae) naturally from non-living matter (organic compound, water, H_2O) at least some 3.5 billion years ago due to global greenhouse effect as mentioned in Chap. 1 (Sect. 1.5).

Tropical rainforests in Andaman and Nicobar Islands in India is a a rich biosphere as shown in Fig. 6.1e. It has about 22,000 varieties of plant species which receive an average total of 2,000 mm of rainfall throughout the year.

(g) **Global ecology**: It deals with study of the Earth's ecosystems with interactions among land, atmosphere and oceans. It helps to understand the large-scale interactions and their influence on the planet Earth with four major environmental issues, namely habitat destruction, invasive species, decline of population densities and pollution.

(h) **Landscape ecology**: It is the science to study and deal with the exchange of energy, materials, organisms and other products of ecosystems. In other words, improving a relationship between ecological processes in the environment and particular ecosystems. Landscape ecology describe the role of human impacts on the landscape structures and functions.

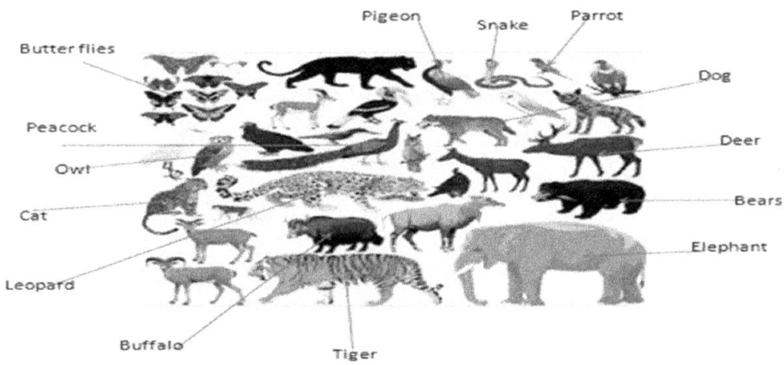

Fig. 6.2 View of wildlife

(i) **Organism Ecology**: Organism ecology is the study of an individual organism's behavior, morphology, physiology, etc. for interaction with biotic and abiotic components in response to environmental challenges.

(j) **Molecular Ecology**: The study of ecology focuses on the production of proteins. These proteins affect the organisms and their surrounding environment. This happens generally at the molecular level.

Deoxyribonucleic acid (DNA) forms the proteins which interact with each other and the environment. These interactions give raise to some complex other organisms.

Energy, ecology and environment have a strong binding to sustain the Nature. Figure 6.1 shows the theme of energy, ecology and environment which comes from nature. In the broadest sense, Nature is the phenomena of the physical world or universe [1]. The study of nature is a very large part of science and humans are part of nature. The nature is generally refers to (i) **Geology**: It is also known as 'geosciences' or 'Earth science'. The study of the solid and liquid matter which constitutes the Earth and (ii) **Wildlife**: It refers to undomesticated animal species which include all plants, micro-organisms animals namely elephant, tiger, snake, leopard, parrot, peacock, buffalo (flora and fauna, Van jiv samuh) etc. It grows or lives in wild areas without interference from humans as shown in Fig. 6.2. Human activity is generally considered a separate category from other natural phenomena [2].

6.3 Geology

Geology describes the structure of the Earth above and below the surface. Further, it provides tools to determine the relative and absolute ages of rocks found in a given location (see Fig. 1.3). It also describes the histories of rocks found above the Earth [3] and demonstrates the age of the Earth. Geology has also provided the primary evidence for (i) plate tectonics theory, (ii) the evolutionary history of life, and (iii) the Earth's past climates.

The plate tectonics theory explains explained as follows:

(a) Earth's outermost layer (lithosphere) as shown in Fig. 6.4 which is made up of the crust and upper mantle is broken into a large number of rocky plates. These are known as tectonic plates.
(b) These plates move relative to each other from 5 to 15 cm (one inch to six inches) per year at different rates,

These plates lie on top of a partially molten layer of rock known as the asthenosphere (see Fig. 6.5). The interaction between tectonic plates is responsible for different geological formations such as the Himalaya mountain range in Asia (see Fig. 6.3), the East African Rift, and the San Andreas Fault in California (USA).

Geologists have been using a wide variety of methods to understand the Earth's structure and evolution, such as field work, rock description, chemical analysis, physical experiments, and numerical modeling. It is important for mineral and hydrocarbon exploration and exploitation, evaluating water resources, understanding natural hazards, remediation of environmental problems etc. It also provides insights into past climate change.

Fig. 6.3 Top view of Himalayas

6.4 Wildlife

Wildlife has been generally found in all ecosystems including deserts, plains, grasslands, woodlands, forests, tropical rainforests and developed urban areas. Some wildlife threatens human safety, health, property, and quality of life. However, many wild animals, even dangerous ones, have value to human beings. This value might be economic, educational, or emotional. Hence, there is a need for conservation of wildlife.

Conservation of wildlife involves the importance of nature which created all living organisms. It is important to protect (a) the endangered plants and animal species and (b) their natural habitats existing on Earth. It will help future generations of wildlife and humans to enjoy.

The following steps are recommended for conservation of wildlife (Source: www.aarushishonik23.files.wordpress.com)

1. There should be the laws made by each government to ban the killing/capturing of endangered animals/birds making these punishable offenses. The laws should be applied strictly without favor and not remain on paper only.
2. The forest authorities should be responsible for implementing the approved laws on wildlife to avoid indiscriminate killing of wild birds/animals.
3. In order to conserve wildlife, each government should establish National Parks and Sanctuaries for preserving the natural habitats of wild animals and birds throughout the country.
4. There should be periodic surveys in all the forests for survival/demise of wildlife (population of all the species of wild animals and birds) for statistical analysis showing growth and decline over the years.
5. All should be made aware of to the conservation of endangered species of wild animals and birds to avoid their extinction.
6. The illegal cutting of living forest trees for timber to build houses and wood for fuel for domestic purposes should be stopped immediately. The living trees are a good absorber of carbon dioxide (CO_2) from the environment and supply oxygen (O_2) to it.
7. If there is a strong need to cut the down trees for development work, then an equal number of tree should be planted nearby.
8. It is the duty of us all to plant trees surrounding homes and nearby parks and encourage others to do the same.
9. The breeding programs through a group meeting for endangered species should be organized locally by forest officials and non-government organizations (NGO).

Humans have historically tended to separate human civilization from wildlife in a number of ways which includes legal action, social responsibility, and moral value/ senses. Animals such as bears, lions, hedgehogs, crocodiles, monkeys, pandas, deer, black leopards, tigers, cheetahs, zebras, jaguars, hippopotamuses, wolves, elephants, chimpanzees, gorillas, snakes, eagles and beavers threaten human life/ safety, health, property, and quality of life. But all have value for human beings, in terms of economy, education or emotional content in nature. In modern times, the natural environment has provoked wildlife activists to protest the exploitation of wildlife for human benefit or entertainment.

In order to predict the population growth with increasing years of various species, it is difficult for any animals apart from humans. Trends are for 2,531 monitored populations, following number of different species was conducted [4–6]:

(i) 465 bird species
(ii) 263 monitored populations of 45 terrestrial mammal species
(iii) 195 monitored populations of 37 aquatic mammal species
(iv) 232 populations of 10 reptile species
(v) 148 monitored populations of 26 fish species

It has been observed that global wildlife populations have decreased by 68% between 1970 and now (see Fig. 6.4a) due to human activity, particularly overconsumption, population growth, trade of wildlife and intensive farming. This finding is according to a 2020 World Wildlife Fund's Living Planet Report and the Zoological Society of London's Living Planet Index measure [7, 8].

6.5 Living Planet Index (LPI)

Further, the living planet index (LPI) has been used to predict their increase/decrease in population with increase of years for sake of comparison and hence it is defined as follows:

$$\text{Living Planet Index } (\text{LPI})_{\text{year}} \text{for a given year} = \frac{\text{Population of } \text{n}_\text{i}^{\text{th}} \text{species in the given year}}{\text{Population of } \text{n}_\text{i}^{\text{th}} \text{species in the year 1970}} \tag{6.2a}$$

or,

$$(LPI_i)_{year} = \frac{\left(n_i^{th}\right)_{year}}{\left(n_i^{th}\right)_{1970}} \tag{6.2b}$$

Here, i is number of species available on the planet earth.

Equation 6.2a, b can be discussed as follows:

Case (i): If $\left(n_i^{th}\right)_{year} > \left(n_i^{th}\right)_{1970} \cong (LPI_i)_{year} > 1$ this means population of n_i^{th} species increases with respect to its number at the year 1970.

Case (ii) If $\left(n_i^{th}\right)_{year} = \left(n_i^{th}\right)_{1970} \cong (LPI_i)_{year} = 1$, this means there is no change in population of n_i^{th} species with respect to its number at the year 1970.

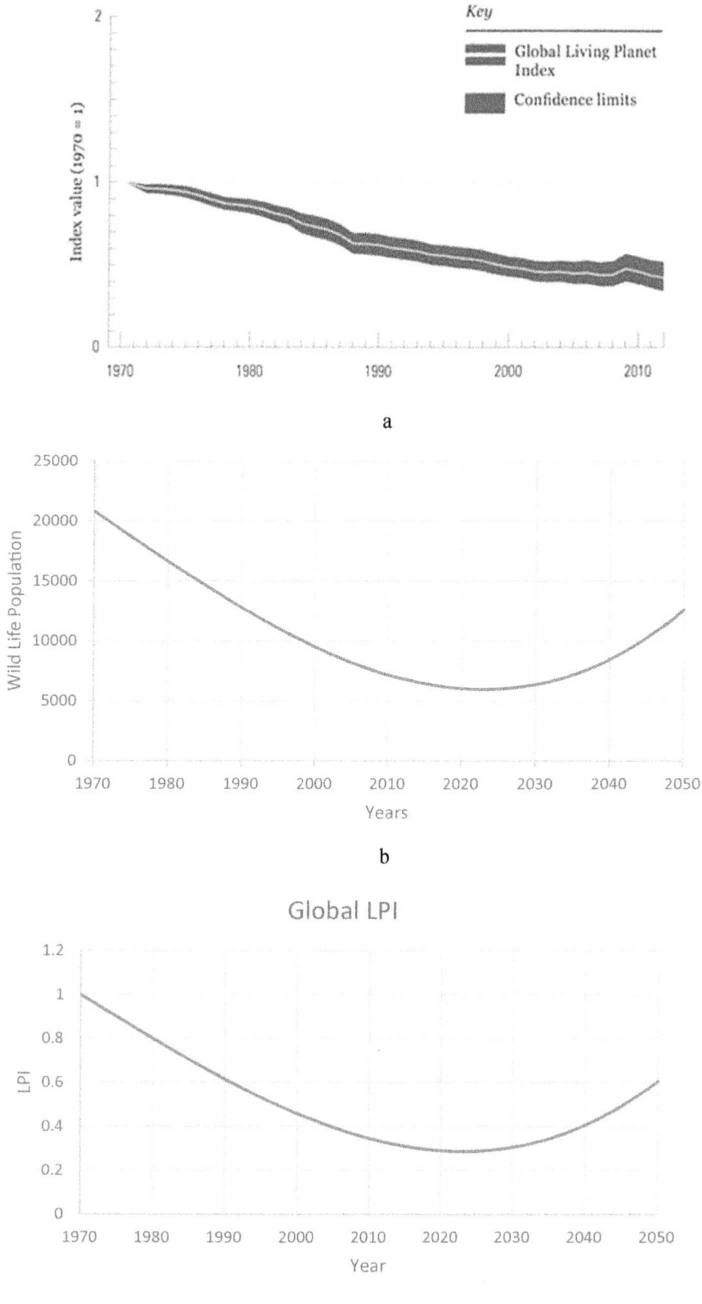

Fig. 6.4 **a** Decline of wildlife population between 1970 and 2010 [6]. **b** Projection of population of wildlife from 1970 to 2050. **c** Variation global living planet index (LPI_1) (wildlife population) from 1970 to 2050

Case (iii): If $\left(n_i^{th}\right)_{year} < \left(n_i^{th}\right)_{1970} \cong (LPI_i)_{year} < 1$, this means population of n_i^{th} species decreases with respect to its number at the year 1970.

Based on findings in Fig. 6.4a, a projection of global population and living planet index (LPI) for wildlife has been developed and its variation with years has been shown in Fig. 6.4b and c respectively.

This indicates that there is decline of global population and LPI of wildlife until 2025, in case (iii), and starts increasing till the year 2050, in case (i). This is a good indication for how to sustain nature. Further, there is increase in population and global LPI for human beings as shown in Fig. 6.5 until 2050, in case (i). However, one can observe that the rate of increase of population and LPI for humans is decreasing until the year 2050 (Fig. 6.5a and b) which may be the reason for increases of wildlife after the year 2025 as shown in Fig. 6.4b. Figures 6.4 and 6.5 has been generated by regression analysis and using the data of ZSL Institute of technology [4–6].

Figure 6.5c shows the yearly variation of living planet index for human, wildlife and its average value. It is seen that there is increase in average LPI for human and wildlife after the year 2000 because of its inter dependent. There are many countries taking care of controlling population which affects the population of wildlife in general. However, global LPI for both human and wildlife indicates an improvement after the year 2030 as shown in Fig. 6.5c.

Further, there is a global decline of tropical rainforest in millions of acres from 1970 to 2030 in case (iii), but this starts increasing until the year 2050 in case (i), and shown in Fig. 6.6. This is also in accordance to Fig. 6.6 due to decreases in the rate of population growth as well increasing plantation due to various government policies, particularly developing and under-developing countries. This is one of the main reasons to sustain and conserve nature. In all cases, solar radiation plays an important role to provide vitamin D to humans as well as wildlife, and photosynthesis to plants to avoid decline of tropical rainforests (Fig. 6.6a–c).

The energy is a basic need of all living organism which play an important role to sustain nature.

6.6 Role of Energy in Ecology

Energy is the capacity for doing work, which can exist in the form of potential, kinetic, thermal, electrical, chemical and nuclear, amongst various other forms. It can be transferred from one body to another. Energy is required by all living organisms to stay alive, such as how human being get energy from food to function day to day. Human beings release carbon-dioxide (CO_2) to the environment which is then consumed by plants as their food. The plant releases oxygen (O_2) to environment to balance environment through photosynthesis (see Sect. 1.5, of Chap. 1. Thus, humans and plants co-exist for sustaining the eco-system. The measure of energy

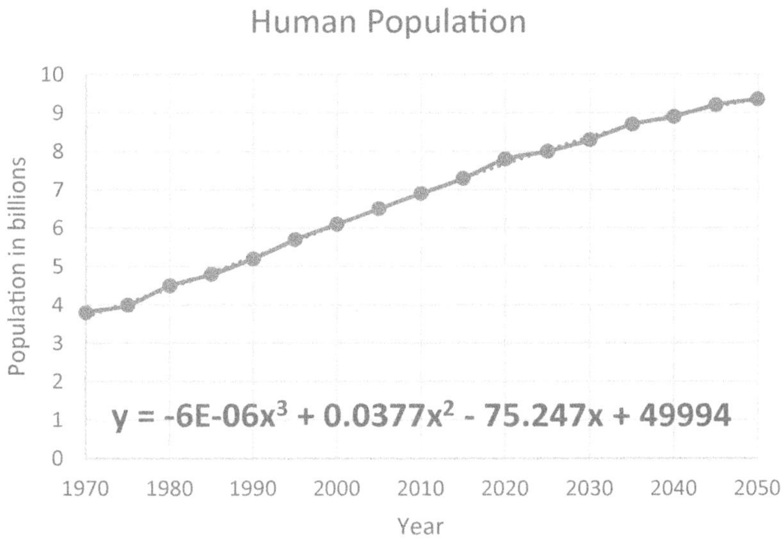

a

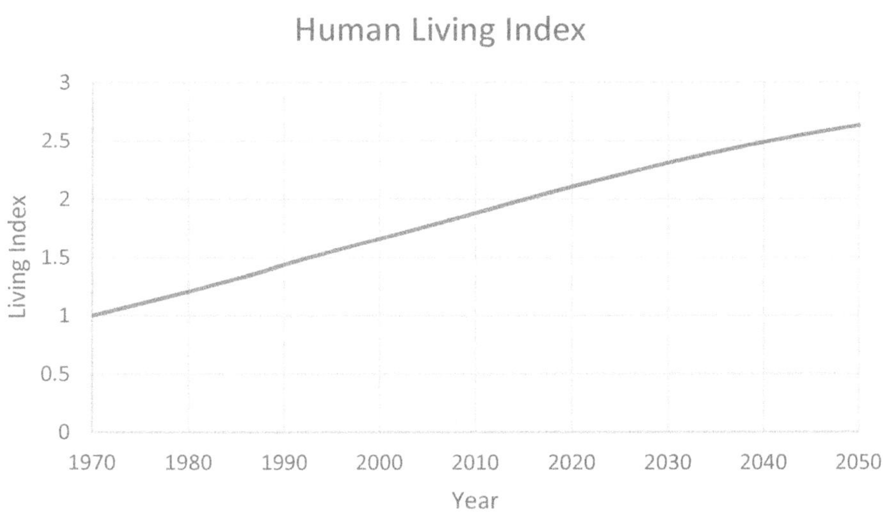

b

Fig. 6.5 **a** Variation of global population, 1970 to 2050. **b** Variation of human living planet index (LPI_2). **c** The variation of living planet index for humans, wildlife and its average. **d** Predicted global LPI including humans

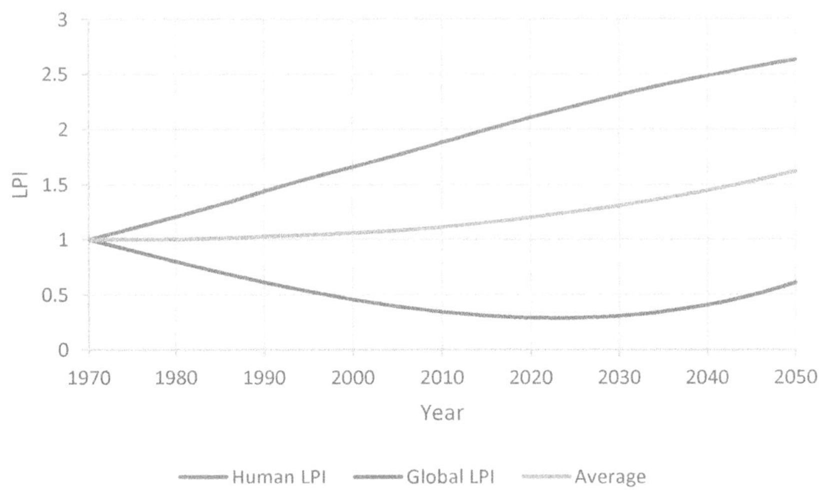

c

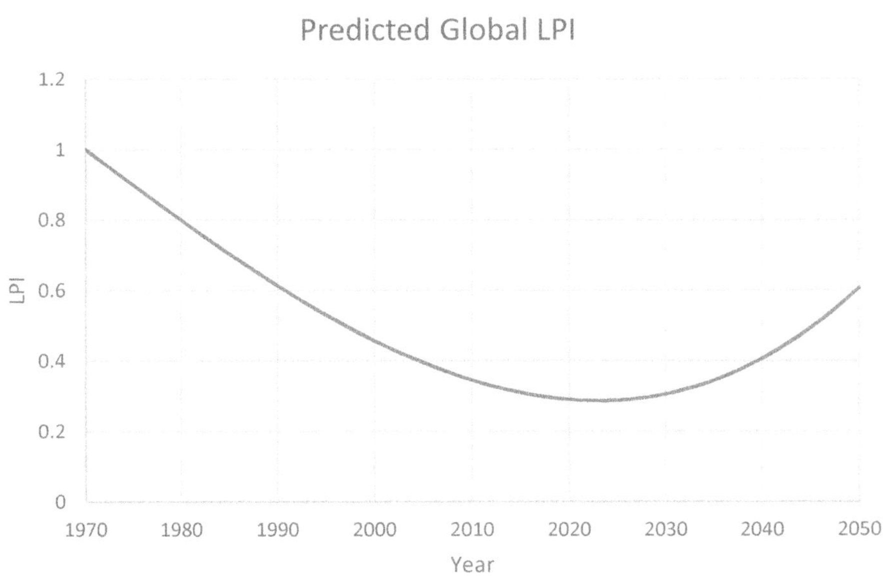

d

Fig. 6.5 (continued)

is in the form of either Joules or Wh in the International System of Units (SI). The energy is also responsible for sustaining nature.

Energy resources can be classified as renewable and non-renewable as mentioned in Sect. 1.1. The basic sources of both energy resources is the Sun either directly (renewable) or indirectly (non-renewable, such as fossil fuels, nuclear fuel), However,

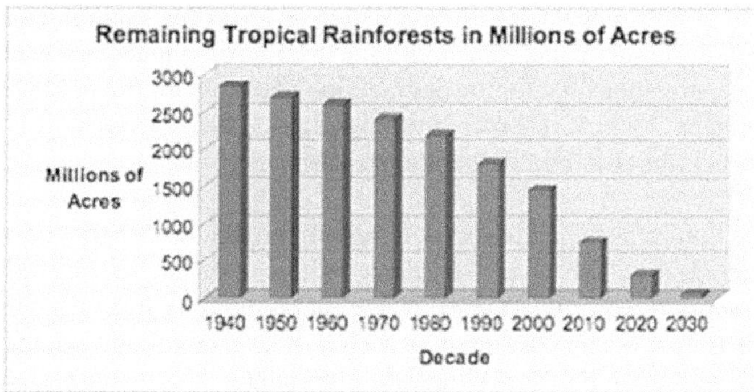

a

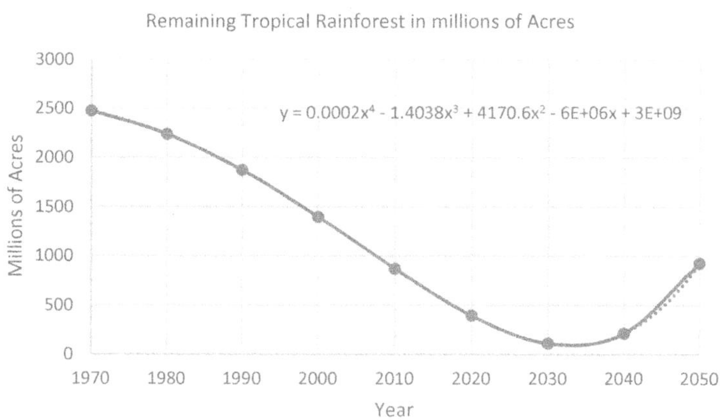

b

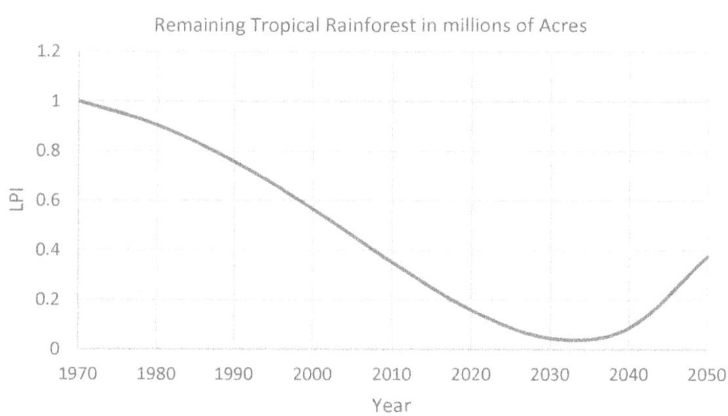

c

Fig. 6.6 **a** Variation of decline of tropical rainforest in millions of acres [9]. **b** Develop correlation for Fig. 6.6a from 1970 to 2050. **c** Variation of LPI_3

the application of non-renewable resources such as fossil fuels and nuclear fuels to meet the needs of human beings through industrial development, particularly after World War Two, is responsible for the degradation of the environment as well climate change, (see Chaps. 4 and 5) respectively. Nowadays, it is preferred to use renewable energy which is infinite in comparison to fossil/nuclear fuels to meet the needs of human beings.

Further, the Sun is the ultimate source of all other renewable energy (wind, hydro, biomass, geothermal etc; see Chap. 3 so it is economical and environmentally friendly to use solar energy directly if possible. Solar energy coming from the Sun through the atmosphere to the planet Earth can be used as both thermal and electrical, which is a basic need of human beings for their development.

Here, it important to mention that non-renewable energy sources, namely fossil fuels (coal, petroleum and natural gases) pollute the environment due to the fast growth of industrialization since World War Two. During electrical power generation from fossil fuels, there is emission of CO_2 and release of thermal energy into the environment. The release of CO_2 during power generation has increased the levels of CO_2 from 273 to 450 ppm which is responsible for the degradation of human health, wildlife and the environment. Further, the release of thermal energy is responsible for increases of environmental temperatures, and finally, melting of icebergs, which affect the climate. Table 5.2 shows CO_2 emission factors from different fossil fuels at source and at delivery point. At delivery point, the value of CO_2 emissions becomes double due to transmission and distribution losses. It would be difficult to return to CO_2 levels in the environment as seen in 1970, because the process is irreversible. But to avoid further increases, one can mitigate CO_2 by using renewable energy sources which are friendly to the environment.

6.7 About the Sun

The Sun is the star at the center of the our Solar System. There is a nearly perfect ball of hot plasma at the core in the center of Sun. The status of sun is in the form of gaseous fluid and it took billion years to be formed. Most of the energy (90%) is generated from fusion reactions ($H_2 + H_2 = He + 25$ meV; see Eq. 1.1) at the core between 0 and 0.23R, where R is the radius. The average black body temperature of the Sun is 6000 K. The heat energy is transferred from core to outer surface of the Sun by convection. The Sun emits radiation by Stefan-Boltzmann's law: namely $E = \epsilon\sigma T^4$.

The physical properties of the Sun includes its radius of about 696,340 km, and the distance from the Earth is 156.06 million km. The inner core temperature is 15 million degrees Celsius, in gaseous state, and the black body temperature of Sun is 6000 K.

6.8 Ecology due to the Existence of the Earth

Earth is the only planet known to support life, and its natural features are the subject of many fields of scientific research. Within the Solar System, it is third closest to the Sun; it is the largest terrestrial planet and the fifth largest overall. Earth is estimated to have formed after the Sun 4.54 billion years ago from the solar nebula, along with the Sun and other planets. It consists of solid particle and has many layers. It can be described by plate tectonic theory as mentioned earlier. The outer layers of Earth are divided into the lithosphere (solid: conduction) and asthenosphere (fluid: convection) (see Fig. 1.4). The lithosphere is cooler and more rigid; while the asthenosphere is hotter and flows. The lithosphere is divided into different moving plates which are considered as the crust.

Earth moves around the Sun in an elliptical path. It takes 365 days to complete one circle. It also spins around its axis to make day and night. One rotation takes about 24 hours. The average temperature of Earth's surface is about 300 K. The heat is transferred from core to outer surface (crust) by conduction.

The temperature of Earth at its core is about 6000K which is similar to the black body temperature of the outer surface of the Sun.

Following Eq. 1.2, if $T = 6000$ K then $\lambda = 0.5$ μm, an average short wave length radiation coming from the Sun and having maximum thermal energy associated around this wavelength and for $T = 300$ K (Earth' surface temperature) then $\lambda = 10$ μm, an average long wavelength emitted from the Earth having minimum thermal energy associated as explained earlier.

Forests absorb CO_2 from the environment on the planet Earth to maintain an optimum level of CO_2 in the environment (see Fig. 6.8) accounting for about one-third (38%) of habitable land area. In other words, this is around one-quarter (26%) of both habitable and uninhabitable land area on the planet earth. At present, global forest area has been reduced by half due to the expansion of agriculture to meet livestock required for the increased population.

$$\textbf{Today, half of global habitable land area} = \textbf{the world's forests area}. \quad (6.3)$$

6.9 Role of the Atmosphere in Ecology

The atmosphere, as shown in Fig. 6.2 is a mixture of greenhouse gases, namely CO, CO_2, O_3, NO_x, SO_x, ozone (O_3), dust, H_2O etc. erupted from within the Earth through cracks between tectonic plates as explained earlier in Sect. 1.1. It surrounds the planet Earth and exists between the Sun and the Earth and it is porous too. On Earth, the atmosphere helps make life possible due to the global greenhouse effect. Besides providing us with something to breathe, it shields us from most of the harmful ultraviolet (UV) radiation coming from the Sun due to presence of the ozone layer. It warms the surface of our planet to about 33 °C (59 °F) via the global greenhouse effect. Without the Ozone layer in the atmosphere, life on planet Earth would have been very difficult. The plants could not survive and grow in the presence of heavy ultraviolet (UV) radiation in the terrestrial region. The same thing happens to plankton (organisms floating in the sea) that serve as food for most of the ocean life. With a weakening of the ozone layer in the atmosphere due to the launch of many satellites, humans would be more susceptible to skin cancer, cataracts and impaired immune systems. The satellites discharge quantum of corbon mono-oxide (CO) during launch in the atmosphere and discharged CO converts ozone into oxygen (O_2), hence the ozone presence in the atmosphere is depleted. Another reason for depletion of ozone in atmosphere is due to human activities such as using chemicals containing chlorine and bromine. These chemicals combine with certain weather conditions to cause reactions in the ozone layer which leads to ozone molecules being destroyed.

The atmosphere largely prevents extreme differences between daytime and night-time temperatures. The other planets in our solar system also have an atmosphere, but none of them have the same ratio of gases and layered structure as Earth's. The atmosphere also acts as a transparent media which allows only solar radiation between wavelength of 0.3–3 μm coming from the sun to reach us, and the rest is either reflected back to the extra-terrestrial region or absorbed. Further, it behaves as an opaque layer for long wavelength radiation ($\geq$ 10 μm) emitted from the Earth.

Without porous atmosphere, all solar energy emitted by the Sun would be reflected; this is known as the albedo effect. With this, the temperature of Earth would fall to around –18 °C (frozen conditions) and the whole Earth would be an iceberg initially.

Finally, the atmosphere consists of various gases transmits short wavelength radiation ($\leq$0.3 μm) but behaves as opaque for long wavelength radiation ($\geq$10 μm).

The atmosphere consists of greenhouse blocks the long wave length radiation emitted by Earth. So the ambient air temperature of environment in terrestrial region starts increasing. If ambient air temperature is higher than iceberg on planet Earth, then ice is melted and it becomes stagnant water. Now, Solar radiation generally considered into two forms:

(a) Electromagnetic (e/m) wave: Its wavelength ranges from 0 to ∞μm and contains thermal energy.
(b) Ray optics: It consists of photons: $E = h\nu$.

If photons ($E = h\nu$) strike water molecules (stagnant water), then H_2O is ionized into H_2 and O_2. During this process, O_2 is released to the atmosphere and H_2 captures carbon di-oxide from the environment to make carbohydrates. Carbohydrate is green algae. Hence the whole phenomenon is known as Global Greenhouse Effect (GGE) (see Chap. 1). This is the main reason for survival of all living organisms including human beings and wildlife.

Fig. 6.7 View of algae formation in stagnant pond water (aquatic ecosystem) due to global greenhouse effect

Algae were most likely the first living plant life on Earth over 3.5 billion years ago, (see Fig. 6.7). These living organisms are not necessarily closely related. However, certain unknown features unite them, while distinguishing them from the other major group of photosynthetic organisms (the land plants).

Global greenhouse effect is responsible for sustain ecological system (ecosystem) in terrestrial region of Fig. 1.2.

6.10 Composition of Environment for Ecology

Dry air in the environment (Earth's atmosphere) is composed of 78% nitrogen (N_2) and about 21% oxygen (O_2). Argon, carbon dioxide (CO_2), and many other gases makes up the remaining 1% of the atmosphere's mixture of gases. The environment also includes water vapor constitutes about 1%. There are also many small particles—solids and liquids—"floating", known as "aerosols" (dust, spores and pollen, salt from sea spray, volcanic ash, smoke etc.).

6.11 Results and Discussion

In Sect. 6.1, we have only considered three species namely (a) wildlife, (b) human and (c) tropical rainforest. However, we have considered seven groups: birds, reptiles, mammals (aquatic), mammals (terrestrial), fish and wildlife as shown in Fig. 6.8a to develop correlation to predict global population growth and LPI. By using regression analysis, we have developed an empirical correlation to predict the population of seven species from 1970 to 2050 as given in Table 6.1. It is based on data's generated by ZSL Institute of technology [6]. Average abundance change between 1970 and 2018 in different taxonomic groups of species listed on Convention on the Conservation of Migratory Species of Wild Animals (CMS). 2018. Appendices I and II of the Convention on the Conservation of Migratory Species of Wild Animals (CMS). https://www.cms.int/en/page/appendix-i-ii-cms (Fig. 6.8b).

Further, global population for remaining five species and LPI for seven species has also been shown in Fig. 6.8a, b. Figure 6.8a has been reproduced by using the regression correlation given in Table 6.1. It is evident from Fig. 6.5a, b and c that the human population is increasing fast, and it may reach 9.5 billion (Fig. 6.5a) with global index of more than 6.5 (Figs. 6.5b and 6.8b). Further, it is to be noted that population and LPI of wildlife is predicted to decrease until 2025 and then increase until 2050 due to precautionary measured announcements by various governments from time to time. However, similar trends have been observed for fish which is the smallest sample considered for the studies in terms of number and LPI. Further, reptiles and mammals (aquatic) have increasing trends, much better than humans, for population and global LPI.

Conservation of total number of living organisms and its population (N) can be expressed as follows:

Table 6.1 Empirical correlation to predict the population of seven species

S.No	Species	Equation
1	Bird	$Y = -0.277211730940781x^2 + 1113.17918561151x - 1114502.06125441$
2	Reptiles	$y = 0.294815027845726x^2 - 1161.56052673357x + 1,144,346.57795535$
3	Mammals (Aquatic)	$y = 0.182278596569693x^2 - 722.345519626859x + 715,829.553271606$
4	Mammals (Terrestrial)	$y = 0.067659223977615x^2 - 269.934855931161x + 269,444.445509642$
5	Fish	$y = 0.0713102782159891x^2 - 287.395272227184x + 289,576.283963176$
6	Human	$y = -4596.72562067769x^3 + 27,554,916.9502577x^2 - 54,974,196,727.2728x + 36,508,672,703,781.5$
7	WildLife	$y = 0.048434143398314x^3 - 286.097915726788x^2 + 562,897.124631225x - 368866215.595019$

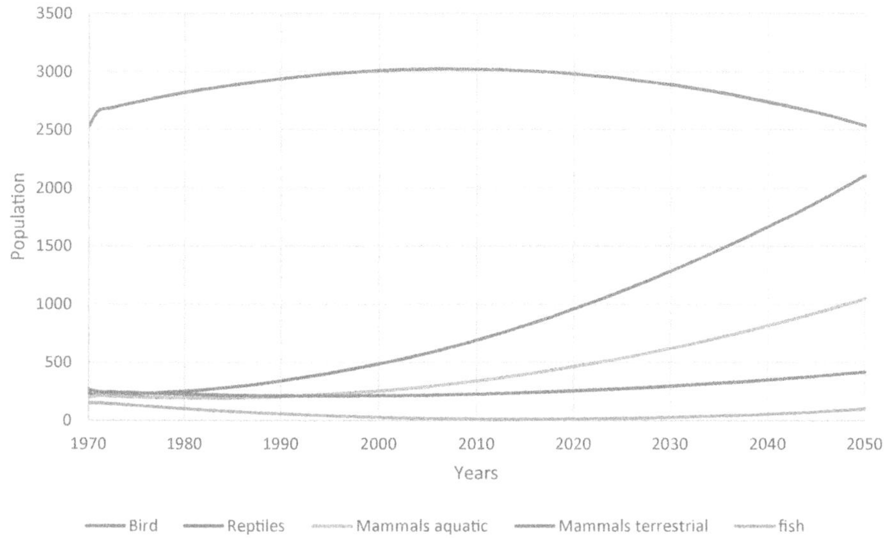

a

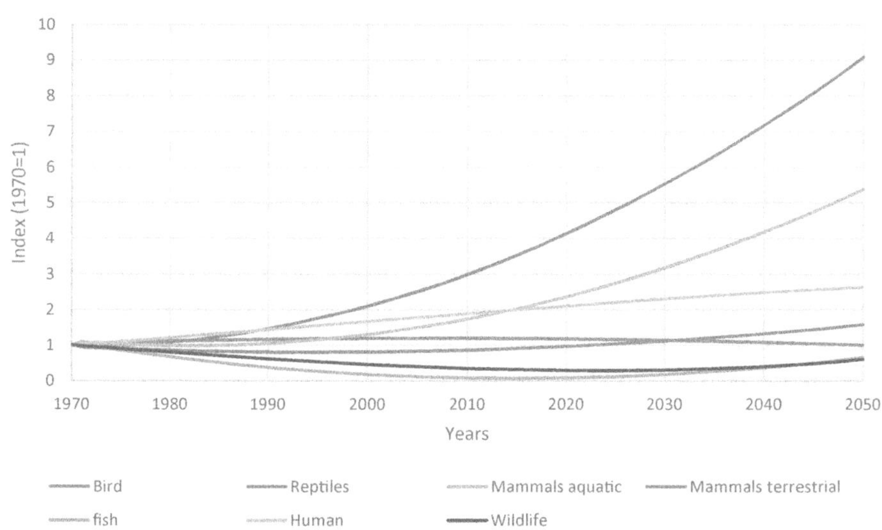

b

Fig. 6.8 **a** Population of five species excluding wildlife (Fig. 6.5a) and human (Fig. 6.6a) from 1970 to 2050. **b** Global LPI variation from 1970 to 2050

$$N = \sum_{j=1}^{n_2}\sum_{i=1}^{n_1} n_{ij} = \sum_{j=1}^{n_2} \eta_{1j} + \sum_{j=1}^{n_2} \eta_{2j} + \sum_{j=1}^{n_2} \eta_{3j} - - - - - - \tag{6.4}$$

where,

i = number of species/communities (8.7million); j = population of species/ communities;

Here, it is important to mention that i and j will change over time. It may either increase (Fig. 6.5 or decrease (Fig. 6.6), or be eliminated from the planet Earth to maintain N as to be constant. This change may be due to human activities as well as weather, environment, eco-system and climate change. The environment and climate change is due to heavy use of fossil fuels, particularly in developed nations (DN).

The organisms (biology) are a discrete and complete living things, such as animals, plants, fungus or microorganisms, and species is a type or kind of thing; i.e. a group of plants or animals having similar appearance.

Here, a total living planet index (TLPI), in addition to global living planet index (LPI), will be defined. For this, one can express mathematically total living planet index (TLPI) after dividing Eq. 5 by N with reference to the year 1970 as

$$TLPI = \frac{N}{N} = 1 = \sum_{j=1}^{n_2} \frac{\eta_{1j}}{N} + \sum_{j=1}^{n_2} \frac{\eta_{2j}}{N} + \sum_{j=1}^{n_2} \frac{\eta_{3j}}{N} - - - - - -$$

or,

$$TLPI = TLPI_{1j} + TLPI_{2j} + TLPI_{3j} - - - - - - - = 1 \tag{6.5}$$

where $TLPI_1$, $TLPI_2$, $TLPI_3$ etc. are total living planet index (TLPI) for each species/ communities which will vary with years but it will be considered as one at the time of 1070, Eq. 6.5. However, it will be always less than one ($TLPI \leq 1$). But sum of all TLPI will be one which is required to sustain the Nature. Further to measure population of each species is very difficult due to migration of some species from one region to other region. However, it is not possible to have TLPI equal to one due to following reasons:

(i) Elimination of some species from earth, particularly by human beings, to meet their food demand.
(ii) The rate of decline of population of individual species due to fighting among themselves outweighing reproduction
(iii) Events such as cloud burst, floods, earthquakes, sudden climate and environment changes etc.
(iv) Viruses such as COVID-19, cholera, infection, lower respiratory, bad quality air and water, drought etc.
(v) Wars and other large scale conflicts.

(vi) Global sudden increase of industry to meet the demand of human being.

However, nowadays, global LPI (Eq. 6.5) will be generally always less than one, except for human beings. If one assumes a reference year of 1970, when Americans celebrated the first Earth Day on April 22, 1970, then LPI for each species should start from 1 and it should either increase or decrease with each year (see Fig. 6.8b). The level of CO_2, the planet's most important greenhouse gas, was measured at that time as 325 part per million (ppm). If it is increasing, it means the population of individual species is increasing, otherwise it will decrease. The increase of LPI is not as bad as a decrease because if LPI is decreasing, it means after some years it will disappear from the planet, earth which is not good for ecological balance. The reasons for decrease of any species may be due to following:

(i) Sudden change in weather condition for each climate.
(ii) Change in yearly climate due to change in local temperature.
(iii) Frequent occurence of diseases in species.
(iv) Human made reasons for fulfilling of their needs such as food habit by killing animals and creating shelter by cutting down trees.
(v) Development of industries in developing as well as under-developing countries with reference to year 1970.
(vi) Imbalance of carbon dioxide (CO_2) and oxygen (O_2) in the atmosphere due to LPI increase/decrease.
(vii) Deforestation due to fire in many countries.

6.12 Recommendations

In order to sustain nature, following measures should be adopted:

(i) Population control over the world by different technique and agencies in each countries so LPI of human can approach one.
(ii) Greed of humans should be minimized.
(iii) Reuse/recycle materials by various means to meet requirements.
(iv) Deforestations should be avoided to retain LPI close to one, and plantation should be increased for balancing the eco-system.
(v) Maximum use of herbal medicine for treatment of all kind of species.
(vi) Energy per capita for each county should be monitored and it should not be wasted to mitigate CO_2 in the environment.
(vii) Ozone (O_3) layer in the atmosphere should be maintained by minimizing the launching of satellites.

Problems

Problem 6.1 Calculate the per capita growth rate and the actual number of the population of 500 buffalo for annual growth and death of 10% and 5% respectively.
Hint: See Example 6.1b.

Problem 6.2 Evaluate population of a city at end of 2nd year for initial population (N_0) of 1 million and annual population decline (r_d) of 5% due to COVID-19.
Hint: Use Example 6.2a, b

Problem 6.3 Evaluate population of a city at end of 5th year for initial population (N_0) of 1 million and first two annual population decline (r_d) of 5% due to COVID-19 and remaining three year annual population growth of 5%.
Hint: See Examples 6.1a, b and 6.2a, b.

Problem 6.4 Derive time period (t) if the population of city is increased from one million to ten million if annual population growth is 5%.
Hint: Example 6.3.

Problem 6.5 Derive for annual population decline in terms of carrying capacity (C) and discuss limiting cases.
Hint: See Example 6.3.

Problem 6.6 Calculate sample standard deviation for the following data of cows in villages.

	60	75	110	130	150	165	195	225	220
Cow	2010	2011	2012	2013	2014	2015	2016	2017	2018

Hint: Here, n = 9, consider 2010 as 1, and Use Eq. 6.5.2.

Problem 6.7 Calculate sample standard deviation for the following data of villagers.

Villagers	5000	5250	5500	5720	6000	6300	6430	6800	7000
Year	2010	2011	2012	2013	2014	2015	2016	2017	2018

Hint: Here, n = 9, consider 2010 as 1, and use Eq. 6.5.2.

Problem 6.8 Calculate covariance and correlation of Problems 6.5 and 6.6
Hint: Use Eqs. 6.5 and 6.6.

Problem 6.9 Write down an expression for effective growth of population b_{eff} by considering immigration effect $I(t)$.
Hint: Use Example 6.1a, b for $b_{eff} = \frac{[B(t)+I(t)]}{N_0}$.

Problem 6.10 Write down an expression for effect on population d_{eff} by considering immigration effect $I(t)$.
Hint: Use Example 6.1a, b for $d_{eff} = \frac{[D(t)+E(t)]}{N_0}$.

Problem 6.11 Write down the population growth/decline rate per capita (r) with immigration and emigration.
Hint: Eqs. 6.1.1 and 6.1.2.

Problem 6.12 Derive an expression for population after time 't' for Problems 6.10 and 6.11 with and without carrying population.
Hint: See Examples 6.4 and 6.5.

Problem 6.13 In the first year, 60 fish are born; if 40 and 20 survived in the second and third years, then what is the mortality rate between the second and third years?
Hint: Example 6.1a, b.

Objective Questions

6.1 The population growth rate per capita (r_g) without immigration and emigration can be evaluated

(a) $\frac{[B(t)+D(t)]}{N_0}$
(b) $(b - d)$
(c) $(b + d)$
(d) $\frac{[B(t)-D(t)]}{N_0}$

Answer: (b) and (d)

6.2 An expression for the population decline rate per capita (r_d) will be obtained from

(a) $-\frac{[B(t)-D(t)]}{N_0}$
(b) $-(b - d)$
(c) $-(d - b)$
(d) $-\frac{[D(t)-B(t)]}{N_0}$

Answer: (c) and (d)

6.3 For annual population growth rate per capita (r) will be positive for

(a) $(b > d) - \frac{[B(t)-D(t)]}{N_0}$
(b) $-(b - d)$
(c) $-(d - b)$
(d) None of these

Answer; (a)

6.4 The abiotic species are
(a) Water
(b) Air
(c) Cow
(d) Human

Answer: (a) and (b)

6.5 The biotic species are
(a) Water
(b) Air
(c) Cow
(d) Human

Answer: (c) and (d)

6.6 For strong strength of the relationship between variables species, correlation coefficient should be
(a) Equal to one
(b) More than one
(c) Less than one
(d) Equal to zero

Answer: (a)

6.7 For increasing population of any species, iving planet index $(LPI)_{year}$ should be
(a) Equal to one
(b) Less than one
(c) Equal to zero
(d) More than one

Answer: (d)

6.8 For decreasing population of any species, iving planet index $(LPI)_{year}$ should be
(a) Equal to one
(b) Less than one
(c) Equal to zero
(d) More than one

Answer: (b)

6.9 The biosphere consists of
(a) Water and plant
(b) Water and animals
(c) Plant and animal
(d) Flora and fauna

References

1. Nature—Wikipedia. https://en.wikipedia.org/wiki/Nature › wiki › Nature
2. F. Ducarme, D. Couvet, What does 'nature' mean? Palgrave Commun. Springer Nature **6**(14) (2020). https://doi.org/10.1057/s41599-020-0390-y
3. H.R. von Gunten, Radioactivity: a tool to explore the past(PDF). Radiochimica Acta 70–71(s1), 305–413 (1995). https://doi.org/10.1524/ract.1995.7071.special-issue.305. ISSN 2193-3405. S2CID 100441969. Archived (PDF) from the original on 2019-12-12. Accessed 29 June 2019
4. Convention on the Conservation of Migratory Species of Wild Animals (CMS), Appendices I and II of the Convention on the Conservation of Migratory Species of Wild Animals (CMS) (2018). https://www.cms.int/en/page/appendix-i-ii-cms
5. P. Greenfield, Humans exploiting and destroying nature on unprecedented scale—report. The Guardian (2020). Accessed 10 Sept 2020
6. A. Woodyatt, Human activity has wiped out two-thirds of world's wildlife since 1970, landmark report says. CNN (2020). Accessed 10 Sept 2020
7. The Living Planet Index (LPI) for species listed on the CMS Appendices, ZSL Institute of technology, UNEP/CMS/COP13/Doc.24/Annex 5
8. WWF, in Almond, R.E.A., Grooten, M., Petersen, T. (eds.), *Living Planet Report 2020 - Bending the Curve of Biodiversity Loss*. WWF, Gland, Switzerland (2020)
9. Tropical rainforest Deforestation Report, https://hlh.co/deforestation.html

Other References

10. Ecology and Biodiversity—Eden District Council
11. Biodiversity - Wikipedia
12. Biosphere – Wikipedia

Chapter 7
Sustainable Environment

7.1 Introduction

The **natural environment** is a **natural world** which depends on the Sun, the atmosphere and ecology on the planet Earth. It took billions of years to form. The ecology consists mainly of all *living* (biotic) and non-living (abiotic) things occurring naturally, and not artificially. In previous chapters, we have discussed all about in details. The natural environment means the interaction between all living *species* (resources like plants/aquatic and non aquatic animals), *climate* change, weather condition and natural resources for survival of human with economical benefits/activity. A couple of decades back, one had to use the terms air, water, food and energy to refer to the basic needs of humans. Now the term 'clean and organic' has been added to air, water, food and energy due to greed. As mentioned earlier in various chapters, the environment has been polluted due to industrialization since World War Two.

In order to sustain the environment at present levels, we require long-term strategies for energy-related issues. In this regard, the use of renewable energy resources (Chap. 3) are likely to be one of the most efficient and effective solutions. Renewable technologies have been considered to be substantially safer. It can be a solution to many environmental and social problems associated with conventional fossil fuels (Chap. 4).

There have been much active worldwide research and development in the field renewable energy resources and technologies since the oil crisis in the year 1970. Renewable energy conversion technologies appeared to be most attractive due to the cost effectiveness and easy implementation, even in rural areas of developing and under-developing countries. Renewable energy technologies produce marketable energy by converting natural resources like solar energy, biomass, wind and hydro in different part of any countries into useful energy forms [1]. These technologies use solar energy directly or indirectly for a continuous stream of power. It warms us, causes crops to grow via photosynthesis, heats the land and sea differentially which causes winds and consequently waves and, of course, rain, leading to hydropower (Chap. 2). Tidal rise and fall is the result of the gravitational pull of the Moon

G. N. Tiwari, *Energy, Ecology and Environment*,
https://doi.org/10.1007/978-981-99-3997-8_7

Table 7.1 Water distribution on earth by volume (Km^3) and percentage (%)

S.no.	Water source on earth	Water volume (Km^3)	Percent of total water (%)
1	Oceans, seas, and bays	1,338,000,000	96.54
2	Ice caps, glaciers, and permanent snow	24,064,000	1.74
3	Groundwater	23,400,000	1.69
4	*Fresh*	*10,530,000*	0.76
5	*Saline*	*2,870,000*	0.93
6	Soil moisture	16,500	0.003
7	Ground ice and permafrost	300,000	0.022
8	Lakes	176,400	0.013
9	Atmosphere	12,900	0.001
10	Swamp water	11,470	0.0008
11	Rivers	2,120	0.0002
12	Biological water	1,120	0.0001
	Total water	1,399,560,010	$96.54 + 5.16 \approx 100$

Source Igor Shiklomanov's chapter "World fresh water resources" in Peter H. Gleick (editor), 1993, Water in Crisis: A Guide to the World's Fresh Water Resources (Oxford University Press, New York)

Table 7.2 Water net gain due to precipitation and evaporation on the planet Earth

S.no.	Reasons/causes	Oceans (km^3/yr)	Continents (km^3/yr)	Whole earth (km^3/yr)
1	Precipitation	324,000	99,000	423,000
2	Evaporation	361,000	62,000	423,000
3	Net gain	37,000	37,000	0

This shows that storage capacity of the continent is more than the ocean due to hard surfaces.

The conversion of ppb into $\mu g/m^3$ can be obtained as follows:

$$\text{Y concentration}\left(\text{in}\,\mu g/m^3\right) = \\ = \frac{\text{molecular weight of substance}\left(\frac{g}{mol}\right) \times \text{concentration (ppb)}}{22.45}$$

$$= \frac{\text{molecular weight of a substance}\left(\frac{g}{mol}\right) \times \text{concentration(ppm)}}{22.45} \times 10^{-3} \quad (7.1a)$$

or,

$$\text{Y concentration(in mg/m}^3) = \frac{\text{(molecular weight of substance)} \times \text{X ppm}}{24.45} \quad (7.1b)$$

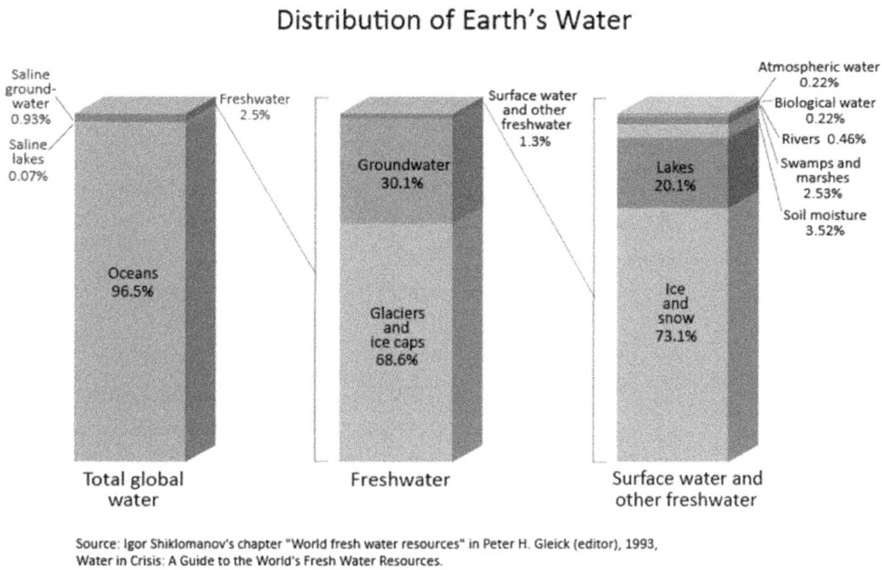

Fig. 7.1 Graphical distribution of water on the planet earth

The conversion of $\mu g/m^3$ to ppb can be obtained as follows:

$$\text{X concentration(in ppb)} = \frac{24.45 \times \text{concentration}(\mu g/m^3)}{\text{molecular weight}} \tag{7.2a}$$

or,

$$\text{X concentration(in ppm)} = \frac{24.45 \times \text{concentration}(mg/m^3)}{\text{molecular weight}} \tag{7.2b}$$

where the number 24.45 in Eqs. (7.1a, 7.1b) and (7.2a, 7.2b) is the volume (litres) of a mole (gram molecular weight) of a gas at the temperature of 25 °C and the pressure of 1 atmosphere (1 atmospheric = 1.01325 bar).

Example 7.2 Compute carbon mono-oxide (CO) concentration in micro-gram per m^3 ($\mu g/m^3$) for a given 9 ppm of CO in air at 25 °C and 1 atmospheric pressure.

Solution

The relation between $\mu g/m^3$ and ppm of a gas is given by Eq. (7.1b)

$$\begin{aligned}\text{Y concentration(in } \mu g/m^3) &= \frac{\text{molecular weight} \times \text{concentration (ppm)} \times 10^{-3}}{22.45} \\ &= \frac{28 \times 0.1}{22.45} = 0.125\,\mu g/m^3\end{aligned}$$

Table 7.3 Concentration of various atmospheric gases for clean and dry air at ground level on the planet earth, 1 ppm = 1,000 μg per cubic meter (mg/m^3)

Gas	Concentration, ppm by volume (part per million)	Concentration, % by volume	Estimated residence time
Nitrogen (N_2)	280,000	78.09	Continuous
Oxygen (O_2)	209,500	20.95	Do
Argon (Ar)	9,300	0.93	Do
Carbon dioxide (CO_2)	320	0.032	2–4 years
Neon	18	0.0018	Continuous
Helium (He)	5.2	0.00052	2 million years
Methane (CH_4)	1.5	0.00015	4–7 years
Krypton (Kr)	1	0.0001	Continuous
Hydrogen (H_2)	0.5	0.00005	–
Nitrogen oxide (N_2O)	0.2	0.00002	4 years
Carbon monoxide (CO)	0.1	0.00001	0.5 years
Zenon (Xe)	0.08	0.000008	Continuous
Ozone (O_3)	0.02	0.000002	~60 days
Ammonia (NH_3)	0.006	0.0000006	7 days
Nitrogen dioxide (NO_2)	0.001	0.0000001	–
Nitric oxide (NO)	0.0006	0.00000006	5 days
Sulfur dioxide (SO_2)	0.0002	0.00000002	4 days
Hydrogen sulfide (H_2S)	0.0002	0.00000002	2 days

Here, gram molecular mass of CO = 12 + 16 = 28 g/mol

Volume of CO gas per mol at 25 °C and 1 atmospheric pressure = 22.45 L/mol.

Volume concentration in ppm = 9 ppm.

After substitution in the above equation

$$\text{Y concentration(in } \mu g/m^3) = \frac{\frac{28g}{mol} \times 9\,\text{ppm} \times 10^{-3}}{22.4\frac{L}{mol}}$$

$$= \frac{28g \times 9\,\text{ppm} \times 10^{-3}}{22.4\,L}$$

$$= \frac{28g \times 9\,\text{ppm}10^{-3}}{22.45 \times 10^{-3}m^3} = \frac{28 \times 9\mu g}{22.45m^3}$$

Table 7.4 Conversion factor and molecular weight of some pollutant gases (1 ppb = 0.001 ppm) (μg = 0.001 mg)

S. no.	Pollutant gases	Conversion factor (WHO)	Conversion (European commission, EC)	Molecular weight ≈ (g/mol)
1	Ammonia (NH_3)	1 ppb = 0.7 $\mu g/m^3$		7
2	Carbon monoxide (CO)	1 ppb = 1.15 $\mu g/m^3$	1 ppm = 1.1642 mg m^{-3}	28
3	Nitric oxide (NO)	1 ppb = 1.23 $\mu g/m^3$		30
4	Nitrogen dioxide (NO_2)	1 ppb = 1.88 $\mu g/m^3$	1 ppb = 1.9125 $\mu g\ m^{-3}$	46
5	Ozone (O_3)	1 ppb = 1.96 $\mu g/m^3$	1 ppb = 1.9957 $\mu g\ m^{-3}$	48
6	Sulphur dioxide (SO_2)	1 ppb = 2.62 $\mu g/m^3$	1 ppb = 2.6609 $\mu g\ m^{-3}$	64
7	Carbon dioxide (CO_2)			44
	Benzene	1 ppb = 2.21 $\mu g\ m^{-3}$	1 ppb = 2.2452 $\mu g\ m^{-3}$	

$$= 11.224\mu g/m^3 \left[\text{as } 1\,\text{ppm} = \frac{1\mu g}{g}\right]$$

Example 7.3 Evaluate carbon mono-oxide (CO) concentration in ppm for 11.224 $\mu g/m^3$ of carbon mono-oxide (CO) in air at 25 °C and 1 atmospheric pressure.

Solution

From Example 7.2 we have

$$\text{Y concentration}(\text{in } \mu g/m^3) = \frac{\text{molecular weight x X concentration (ppm)} \times 10^{-3}}{22.45 \times \frac{L}{mol}}$$

or,

$$\text{X concentration}(\text{in ppm}) = \frac{11.224\,\mu\frac{g}{m^3} \times 22.45 \times \frac{L}{mol}}{28\frac{g}{mol} \times 10^{-3}} = \frac{11.224\,\mu\frac{g}{m^3} \times 22.45 \times L}{28 \times 10^{-3} g}$$

$$= \frac{11.224\,\mu\frac{g}{m^3} \times 22.45 \times 10^{-3} m^{-3}}{28 \times 10^{-3} g} = 8.999 \approx 9\,\text{ppm}[\text{as } 1\,\text{ppm} = \frac{1\ \mu g}{g}]$$

Example 7.4 Find out carbon dioxide (CO_2) concentration in ppm for a given 20 micro-gram per m^3 ($\mu g/m^3$) of carbon dioxide (CO_2 in air at 25 °C and 1 atmospheric pressure.

Solution

From Eq. (7.2b), we have the following equation

$$\text{X concentration(in ppm)} = \frac{24.45\,\frac{\text{L}}{\text{mol}} \times \text{concentration}\left(\mu\text{g/m}^3\right)}{\text{molecular weight} \times 10^{-3}}$$

Here, gram molecular mass of carbon dioxide (CO_2) = 12 + 32 = 44 g/mol (Table 7.1b).

Concentration of carbon dioxide (CO_2) = 20 μg/m^3

Substitute values in above equation, we get

$$\begin{aligned}\text{X concentration(inppb)} &= \frac{24.45\frac{\text{L}}{\text{mol}} \times 20\left(\mu\text{g/m}^3\right)}{44\text{g/mol} \times 10^{-3}} \\ &= \frac{24.45 \times 10^{-3}\text{m}^5 \times 20\left(\mu\text{g/m}^3\right)}{44 \times 10^{-3}\text{g}} = 11.11\,\text{ppm} \\ &= 11.11 \times 10^3\text{ppb[as 1 ppm} = \frac{1\mu\text{g}}{\text{g}}\end{aligned}$$

7.3 Impacts of Renewable Energy Technologies (RET) on the Environment [1, 2]

It has been stated earlier in Chap. 4 on energy conservation that as per the Kyoto protocol and Paris agreement, the use of renewable energy resources should be encouraged to developing and under-developing countries for carbon dioxide (CO_2) mitigation. However, not all renewable energy sources are pollution-free; some inject pollutants into environment, namely the burning of biomass/biogas/ agricultural waste after harvesting of crops in fields etc. On other hand, the global feeling has always been given attention to the impact of non-renewable energy sources (fossil fuels) on the environment due to high concentration of carbon dioxide injection into environment during electrical power production. In contrast, the renewable energy sources (RES) have a ‘clean’ image for a sustainable environment. The only major exception to this general trend is the large hydropower projects. Our experience has taught us that they can be disastrous for the clean environment due to the damage of natural dams when earthquakes occur. It is understood mini hydel and micro-hydel projects for small electrical/mechanical generation are harmless.

In the next section, we will discuss impacts of renewable energy resources on the clean/natural environment.

7.3.1 Insolation/Irradiation Effect on Environment

As discussed in Chaps. 1 and 2, solar radiation/irradiation/insolation comes from the Sun to planet Earth through the atmosphere. Renewable energy sources (RES) do not affect the environment when used, unlike non-renewable energy sources (fossil fuels), and hence they provides low-grade clean energy. Other renewable energy sources are also derived from the Sun. The solar energy systems also do not generate air pollution in environment during its operation unlike fossil fuel systems. The basic environmental issue with the renewable energy technology (RET) is indirectly associated with its embodied energy which should be recovered as early as possible before dismantling. It is embodied energy and has already injected carbon dioxide (CO_2) into the environment during manufacturing and installation. Direct conversion of solar radiation into useful thermal and electrical (DC) energy can be accomplished in many ways. These are as follows:

7.3.1.1 Solar Thermal System: Land Use and Health Hazards

The solar thermal technologies namely flat plate collector (FPC), evacuated tubular collector (ETC), solar water/air heating, compound parabolic concentrating (CPC) collector etc. operating at medium temperature ranges has no adverse effect on the environment during operation except for maintenance during operation (for details of solar energy technologies consult "**Handbook of Solar Energy, Springer, 2016**"). In the case of integration of hot water/ air or space heating/cooling system etc., with building during construction as shown in Fig. 7.2, no additional land is required. The land is generally required for larger capacity of concentrating thermal power plants. In this case, there is of habitat and changes to the ecosystem. There is also no problem regarding health hazards due to the medium operating temperature range ($\leq$100 °C) like fossil fuel use. In all solar systems, waste materials of each component are recycled and hence contribute less to environmental hazards. The embodied energy for 1 m^2 FPC is 578.5 kWh has a life expectancy of about 10–15 years and energy payback period of approximately 1 year. In roof-integrated FPC, the balance of system (BOS) is negligible.

7.3.1.2 Photovoltaic (PV) System: Land Use and Health Hazards

Further, the photo-voltaic (PV; see Chap. 3), system provides high-grade electrical energy, unlike solar thermal technologies. If a PV system is also integrated to the walls and roof of a house as shown in Fig. 7.3, there is no need for using land, and there are reduced transmission and distribution losses and balance of system (BOS) like in the case of a standalone system. It can be noted that there are two types of PV module: opaque (Fig. 7.3a, b) and semitransparent (Fig. 7.3c). Both have advantages and disadvantages. Embodied energy of I m^2 PV module is about

Fig. 7.2 Flat plate collector (FPC) integrated to roof of a house

1071 kWh with expected life of 25–39 years and payback period which depends on insolation of about 3–5 years. Such a system is known as a building-integrated PV thermal (BiPT) system. In opaque PV integration, it reduces incoming solar radiation through the walls and roof (see Fig. 7.3a, b). In the case of semi-transparent PB module integration, it provides direct heating as well as day lighting. There is a rooftop PV system as well.

For larger capacity installation of PV systems in a MW [grid connected/online mode/net metering system], a larger area of land should be available in the region with high solar radiation/insolation and wind velocity. The land should be barren/ desert land (unfertilized) so that cost of production of electricity should be minimal. In this case, the balance of the system plays an important role. The balance of system (BOS), structural support of a smaller PV system ($\geq$100 kWp), for standalone and rooftop systems is 250 and 500 kWh respectively, in offline mode with provision of

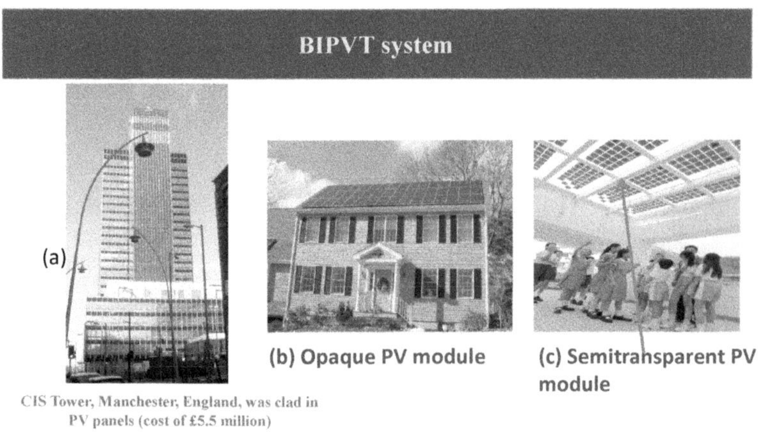

Fig. 7.3 Building integrated PV system without transmission and distribution losses

battery backup. The lands are not usable for agriculture or forest cover. Further, it is required to locate the installed PV systems close to users to avoid transmission and distribution losses, and expenses on installing transmission lines. It is therefore ideal to install PV systems in the arid regions with lower costs of land and high solar intensity with cheap labor to maintain the system.

Figure 7.4 shows an photo of greenhouse integrated semitransparent photovoltaic thermal (GiSPVT) system of capacity 30 kWp located at Margupur, Chilkahar, Ballia (UP), India. The GiSPVT system has been used to generate electrical power and cultivation of various vegetables. This system is most economical for cold climatic conditions (G.N. Tiwari and Arvind Tiwari, Optimization of packing factor for maximum electric power and crop yield in greenhouse integrated semi-transparent photo-voltaic thermal (GiSPVT) system in desert land: An experimental study, *e-Prime—Advances in Electrical Engineering, Electronics and Energy*, Volume 1, 2021, 100,008, https://doi.org/10.1016/j.prime.2021.100008).

Manufacturing solar cells, different materials are used, namely arsenic, cadmium and inert silicon etc. These are hazardous materials that can create health and safety problems for workers coming in contact with them. Workers involved in manufacturing unit of photovoltaic modules and components must seriously be protected from exposure to these materials.

The larger the PV system, the larger the carbon credit (see Chap. 4).

Fig. 7.4 GiSPVT system for pot as well as brick channel cultivation in barren/desert land

7.3.1.3 Solar Thermal Power Plant: Land Use, Construction, Water Resources and Health Hazards

The solar thermal power plant (STPP), generally economical for MW capacity, needs larger land area in arid/desert zone, operates at very high temperature (>500 °C) with use of concentrating collectors connected in series. One Km^2 of land is required for every 20–60 MW. It creates the problem for local desert wildlife protection due to extensive use of mirrors with high reflectivity. In this case, a conventional steam engine is used to generate electricity which requires the use of cooling water. This creates a problem on water resources in arid areas. There is some water pollution due to thermal discharges and release of plant chemicals used in heat exchangers during maintenance of such plants.

The release of heat transfer fluids (water and oil/liquid sodium or molten salt etc.) from concentrating collectors (parabolic trough) and central receiver systems form health hazards that can affect the environment and may be health hazards. A fatal accident could happen in a system using liquid sodium if sufficient care has not been taken. These may be avoided by moving to volumetric systems which use air as a heat transfer medium. Due to high solar intensity concentration from concentrating collectors, there is the possibility of workers losing eyesight around the system.

So far, the solar thermal power plant has not been successful due to high maintenance costs of tracking systems, the discharge of working fluid (pollutants), intermittent solar radiation and high cost of land which cannot be used for other activities as in the case pf PV system (Fig. 7.4). This plant has the usual environmental impacts such as landscape, effects on local ecosystems and habitats, noise and visual intrusion, temporarily pollutant emissions.

Following are additioanl reasons for the failure of the thermal power plant:

(a) It is important to note that it bears substantial costs, hazardous wastes and land-use issues.
(b) Permanent use of large land areas and life of the system.
(c) Recycling of materials after decommissioning like fiber glass, glass, coolant, insulations etc.
(d) High hazard to eyesight from reflectors for workers.
(e) Impact on water resources.

7.3.2 Hydro-Power Plant affects Environment, Fig. 3.7

The basic working principle along with its maximum efficiency with examples has already been discussed briefly in Chap. 3. In this case, water is stored from rainfall in the natural hydro cycle (Chap. 2) and hence it is a renewable source of energy. For installation of a hydropower plant, there is not much land required for waterfall from stored water in a natural dam. However, there is need of reinforced concrete and cement (RCC) structures for the various gates to support the system as shown in Fig. 7.5. However, there is a land requirement for the outflow of water after

Fig. 7.5 Hydro-electric power plant

the power generation stage, which can be used for many purposes like irrigation of agricultural land near by constructed reinforced concrete and cement (RCC) by channels, drinking purposes etc. There are evaporation/respiration losses from water surfaces from stored as well as flowing water through artificial channels. The volume is insignificant in comparison with stored water from rainfall. Efficiency of close to 100% is observed in Pelton wheel turbines due to the curved blade design. India meets about 35–40% of its energy demands from hydropower plants.

These are four important components of hydro power generation system:

(i) **The catchment area**: It is an area over which precipitation takes place. Water stored in a natural dam contains potential energy. The water flows through the catchment area and is referred as "run-off" which can be expressed as

$$\text{Yearly Run} - \text{off} = \text{yearly total precipitation} - \text{yearly total evaporation} \tag{7.3}$$

where evaporation varies with ambient air temperature, wind velocity over the water surface and relative humidity (γ). Run-off increases with low ambient air temperature, low wind velocity over the water surface and high relative humidity due to less evaporation, and vice versa. The larger area of catchment, the larger the run-off.

(ii) **The artificially created lake (reservoir)**: Built by diverting part of a river and storing the water in a reservoir behind an artificially constructed dam to supply water for irrigation and drinking purposes to villagers. In this lake, there is evaporation as well as seepage of stored water. The stored water in the lake is stagnant.

(iii) **Inlet waterway/upstream waterway (Penstock)**: A structure designed to transfer water from intake to the Pelton wheel turbine (see Fig. 3.7a). Further, there is the estuary into which the river flows.

(iv) **Artificial dam and intake house**: The intake house structure is the connection between the headwater (run-off) and Penstock or turbine (see Fig. 3.7a). Generally, there is a screen responsible for keeping floating debris out of the plant at the entrance of the intake.

The following are environmental stresses caused by hydro power plants:

(i) **Timing of river flow variation**: Based on electricity demand (load), there is constant discharge of water flow from the catchment throughout days, weeks and even months. If demand is more in a particular time in a day (peak load) due to increase demand of electricity, constant water flow (run-off) is increased for a certain number of hours. It may also operate in frequency mode, i.e. generators are brought on or off depending on the changing electricity demand throughout the day. Electricity demand is more in the evening than in the day, and in summer. Due to this, there is risk of temperature variation and dissolved oxygen in the downstream water flow. It has been treated as energy storage; it also offers a backup for other intermittent renewable energy sources namely solar energy as well as wind energy.

(ii) **Evapo-transpiration and seepage water losses**: Evapo-transpiration losses are due to evaporation for the hydrologic cycle (Sect. 2.2) and seepage water losses is due to vertically through the bottom of the water pond and horizontally through the dikes by infiltration and through the drainage system of the water pond in catchment area. It depends on porosity of basement of water ponds and cracks, if any. The seepage losses have been given for different materials used for construction of water ponds in Table 7.5. If the catchment surface area is known, then by using the Example 2.1 and multiplying area of catchment, one can get the rate of water lost in kg/s.

Example 7.5 Find water loss from water surface with effective area of 1000 m^2 for Example 2.1 for 24 hours.

Solution

From Example 2.1, we have the rate of evaporation from the water surface as

$$\dot{q}_{ew} = 211.99\text{W/m}^2$$

If latent of heat evaporation of water. $(L) = 2.26 \times 10^6 \text{J/kg}$

Table 7.5 Seepage losses for different natural soil (Estimates of water requirements https://www.fao.org)

S.No.	Natural soil	Seepage (mm/day)
1	Sand	25.00–250
2	Sandy loam	13.00–76
3	Loam	8.00–20
4	Clayey loam	2.50–15
5	Loamy clay	0.25–5
6	Soil with porosity of *0.373*	12,398.4
7	Week weathered rock (foundation)	8.64
8	Sand gravel (lower) (foundation)	12,960
9	Sand gravel (upper) (foundation)	3,456
10	Composite geomembrane (dam)	0.08

The rate of evaporation from the water surface in 24 h (Q_{ew}) will be as

$$Q_{ew} = 211.99W/m^2 \times 24 \times 3066s = 1.8 \times 10^7 j/m^2$$

Then, the total water loss due to evaporation (M_{ev}) can be evaluated as $M_{ew} = \frac{Q_{ew}}{L} \times Area\ of\ water\ surface = \frac{1.8\times10^7}{2.26\times10^6} \times 1000 = 7,964.6kg = 7,964.6l = 7.964m^3$ (see Appendix 1 for conversion of unit).

Example 7.6 Evaluate the amount of water needed in m^3 to compensate for seepage losses through loam surface during 12 months for water pond surface area of 2,000 m^2.

Solution

Seepage losses from loam in one day = 14 mm = 0.014 m/day (Table 7.5, Average of 8–20 mm/day) or,

Seepage losses from loam per day = (0.014 m/day, seepage) × 2, 000 m^2 (pond area) = 28 m^3/day.

Finally, Seepage losses for one year (365 days) = 365 (days) × 28 m^3/day = 10,220 m^3.

(iii) **Barriers to aquatic organism movement**: Due to water flow (run-off), several impacts on the environment such as deterioration of water quality around plants due to different compositions, habitats of various aquatic life and migration barriers and reproduction. Migratory animals like fish (aquatic organisms) need different environments (temperature, oxygen level etc.) at their different phases of life cycle. Different phases are reproduction, nurture of juveniles, growth and sexual maturation. The dam renovation for maintenance after a long time makes it possible for fish migration and fish species to shift from lentic (still waters such as lakes, ponds, or swamps) to lotic (moving water).

(iv) **Greenhouse gas (GHG) emission**: It took place due to construction of plant, the operation and maintenance and the decommissioning of the plant at end of life. The GHG emissions come from the production of the materials by using fossil fuels required for the construction and transport of the work equipment.

(v) **Thermal stratification**: It occurs if two types of stream having different temperatures come into contact. Due to their temperature difference, hot water moves up due to its low density and cold water moves down due to high density (high weight).This affects river water downstream of dams. This also depends on the depths of water columns.

(vi) **Sediment loading and nutrient levels**: Sediment is a thick substance which has been formed at the bottom of a dam/catchment. The problem with removing sediment in the surroundings of the hydropower plant as shown in Fig. 7.6 can be mitigated by either construction of small-scale weirs to trap the sands (sedimentation) or deposited the particles/stone can be manually removed at later stage. A direct approach to reduce the accumulation of sedi-ment at bottom is to mechanically remove the sediments by periodic dredging Following affects have been observed due to sedimentation (a) safety of dams, (b) reduces energy production, storage capacity and, discharge capacity and (c) flood attenuation capabilities. It increases loads on the dam and gates which damages mechanical equipment and creates a wide range of environmental impacts due to the presence of small particulates injected into the environment.

(vii) **Loss of terrestrial habitat to artificial lake habitat**: Due to construction of big dams and reservoirs, environmental splitting (fragmentation) changes the continuous vegetation cover into isolated fragments with the following biological consequences in the terrestrial region:

(a) Forest reduction which affects terrestrial species.
(b) Diversity loss.

Fig. 7.6 Sedimentation surrounding hydropower plant

(c) Shrinking of the habitat area for terrestrial mammals and habitat-specialist species.
(d) Terrestrial and aquatic habitats of a high quality.
(e) Loss of biodiversity due to the restrictions
(f) Edge effect, with sharp cuts in previously continuous forest, resulting in direct influence of wind velocity
(g) Sunlight (solar radiation) and other negative effects on wildlife.
(h) Pushing animal species to the interior of a forest.

(viii) **Population movements**: Large hydropower plants require large reservoirs and discharge areas. Many villages in tribal areas have to be shifted to make room for the installation of a large hydro power plant. This forces the people to live in unprotected areas, who have lived in a relatively small and protected environment. Construction of housing, land distribution by the government, working conditions (new jobs) and ways of life are affected radically. Social consequences are also likely to be affected by dislocating people in ecologically vulnerable areas. The socio-cultural conditions together with their traditional connection to land, water and other natural resources tend to make them unadoptable.

(ix) **Health hazards**: Large hydropower plants can increase the extent of polluted water-related diseases such as cholera, dysentery and several tapeworm and round worm diseaseto people living nearby. The stagnant water reservoir may improve the living and breeding conditions of disease-causing organisms namely pathogens and their intermediate hosts. If the flowing water from the reservoir is employed for irrigation, industrial and drinking water supplies, there may be a the risk of infection by pathogens. Such infection may spread over large areas.

(x) **Flora (all plant life) and fauna (all animal life)**: The changes in flora and fauna in large reservoirs exert a considerable direct effect through submerging the area permanently or periodically.

(xi) **Groundwater**: The filling of a hydro dam reservoir and the flow of a watercourse reduces the groundwater level in surrounding areas. These areas may influence the quality of the underground water and the sediment transport of the watercourse due to run-off and erosion.

(xii) **Excessive fertilization**: There are possibilities to increase the growth of algae or large amounts of higher-order aquatic plants which reduce oxygen levels in the water of the reservoir. It may be due to trapped nutrients. Reduction in oxygen levels can damage aquatic animals. It requires excess biological oxygen. Evaporation also causes a concentration of nutrients leading to excessive fertilization or eutrophication.

(xiii) **Transport of nutrients**: A reservoir in hydro power plants serves as a trapping machine for nutritious elements and mud flowing in. It leads to a reduction of transport of nutrients through downstream. The daily, monthly and annual variations in supply for downstream reduce the biological productions.

7.3.3 *Wind Energy*

Wind energy is one of the intermittent renewable sources of energy as mentioned in Chap. 3 without any air pollution. The advanced three-blade horizontal axis wind turbine is most efficient, as shown in Fig. 7.7. It poses no threat to public safety. It has concerns over the visibility and the noise of the wind turbines and their impacts on the wilderness areas. There are some pros and cons of environmental impacts of wind energy.

7.3.3.1 Advantage (Pros) of Wind Energy Conversion System

(a) **Free fuel (input energy)**: There is no need of fuel due to freely moving wind through the wind cycle (Sect. 3.2.7, Chap. 3). The horizontal wind turbines run on the power of wind-generated kinetic energy, hence there is no need for fuel. Once it is installed, it doesn't require any fuel which reduces the overall cost to operate on large-scale wind farms.

(b) **Clean forms of energy**: Wind energy (KE) does not emit green house gases and hence there is no effect on the environment and climate change. The wind energy only releases greenhouse gases (GHG) indirectly during the manufacturing and transport of the wind turbines material during the installation process due to embodied energy.

(c) **Farmland operations**: After installation of wind energy conversion system (WECS), the remaining land below it can be used for either agricultural purposes or installation of PV systems to make whole system hybrid power generation as shown in Fig. 7.7. This is a great boon to farmers to increase income, due to the minimal space required at ground level.

Fig. 7.7 Wind energy conversion system (WECS)

7.3.3.2 Disadvantages of Wind Energy Conversion System (WECS)

(a) **On life of wildlife**: The moving blades pose a threat to migratory paths of flying birds and bats. In this case, they could be injured or killed, affecting the conservation of their family.

(b) **Noisy condition**: Wind turbines are noisy. Due to this, the WECS is installed in remote areas to avoid disturbance at night. Depending on the location of the turbine, such as offshore/coastal regions, noise may not pose a problem. There are two types of noises: **(i) Mechanical sources**: Such as from the gearbox, generator and auxiliary motors and **(ii) Aerodynamic source**: From the blades passing through the air.

(c) **Expensive cost and embodied energy**: The structures of WECS as shown in Fig. 7.7 are generally more than 30–40 m and require substantial investment due to the large quantity of materials for different corrosion-free components. Further, there are transmission and distribution losses due to the long distance to the users. This requires further investment in underground lines to send power to more populated areas like towns and cities. The majority of the cost is the initial installation and building stage. Maintenance costs are minimal.

(d) **Unpredictable yield**: Wind energy is an intermittent source of energy like solar energy, and varies according to wind velocity. It is hard to predict power yield and hence requires battery storage. This means suppliers and cities need to have energy reserves or alternative sources of power like PV systems as shown in Fig. 7.8. Such systems are referred to as hybrid renewable energy systems. The use of batteries is an expensive phenomenon due to replacement and maintenance costs. In order to avoid the use of batteries (effecting an off grid system) one can have online, net metering or grid connected systems.

(e) **Sustainable energy**: WECS is clean energy and friendly to the environment and climate. It is sustainable sources like hydro, biomass and solar energy.

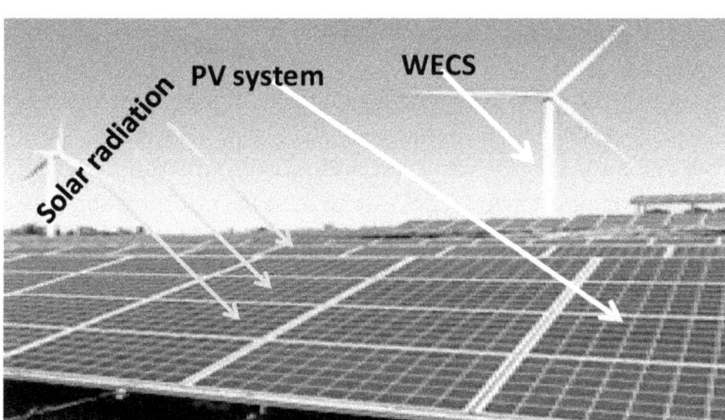

Fig. 7.8 Hybrid renewable energy system (wind and PV)

(f) **Electromagnetic interference**: Wind turbines can interfere with television and microwave signals through reflection which is caused by the rotor blades of turbines intercepting the television beams. The tower structure also scatters the beam. This can be easily solved by careful siting and modification of transmitters and receivers.

(g) **Human safety hazards**: Large wind turbines are generally located in sparsely populated areas so the risk of human accident is minimal.

7.3.4 Biomass Energy

Biomass is plant-based material namely wood, wood residues, *energy crops*, agricultural residues etc. obtained by photosynthesis with the help of solar energy. This is known as light reaction (Eq. 1.4) and its mass increases in the night (night reaction). During photosynthesis, it serves as a sink for atmospheric CO_2 from the environment, and also contributes to soil fertilization and soil stabilization. Hence, it helps to reduce carbon dioxide (CO_2) levels in the atmosphere and minimizes acid rain.

It is used as a *fuel* either for *heat* (thermal low-grade energy) or *electricity* (high-grade energy) production. The caloric value and combustion efficiency of some biomass has been given in Table 7.6. It is the *mass* of living biological *organisms* in a given area/*ecosystem* at a given time and sometimes is referred as bio-energy. The total living biomass/bio-energy on the planet *Earth* is about 550–560 billion tons of carbon (C). The total annual *primary (basic) production* of biomass for converting into energy is just over 100 billion tons of carbon (C) per year. The total living biomass of *bacteria* may be equal to various plants and animals. Biomass has both favorable and unfavorable (adverse) effects on environment due to presence of carbon in it and it acts as a source of renewable energy due to its direct dependency on solar energy. Biomass/bio-energy also improves in the physical properties of the soil along with reduction of water run-off and desertification. It is economical and an inexpensive source of energy for people living in the rural area of developing and under-developing countries.

Table 7.7 shows that there are about 19.4 and 16.7 Megatons of carbon dioxide emitted (CO_2) in to environment in the year 2020 and 2019 respectively with indication of 16% increase in CO_2 emission and mostly emission was due to energy supply and management of water and waste (due to the co-firing and co-gasification of biomass in power plants). There was decrease from 15.1 to 13.7 Mega tons of emissions of CO_2 from 2012 to 2016 due to biomass burning have gone down.

Example 7.7 Evaluate the average solar energy available per unit area per unit time in cal/min.cm^2 in extraterrestrial region.

Solution

The solar constant (I_{sc}) which is measured in extraterrestrial region and its value is 1367 W/m^2 (Example 1.4). Hence,

Table 7.6 Caloric value of some biomass

S.No,	Biomass with condition	Combustion value/caloric value, kcal/kg (CO_V)	Combustion value/caloric value, kJ/kg (CO_V)	Burning efficiency (%) (η_b)
1	Wet wood	2,220	9,288.50	30
2	Dry wood	**3,890**	15,886.76	60
3	Straw	**3,346**	14,000–16,000	60
4	Coal coke	**6,931**	29,000	70
5	Cow dung cake (open)	2,092	8,753	11
6	Gobar gas	4,713 per m^3	19,719	60

Table 7.7 CO_2 emission by combustion of biomass (Carbon emissions from burning biomass for energy· https://www.pfpi.net› uploads› 2011/04)

S. no.	Year	CO_2 emission, mega tons
1	2011	14.8
2	2012	15.1
3	2013	14.1
4	2014	13.8
5	2015	13.7
6	2016	13.7
7	2017	13.9
8	2018	14.7
9	2019	16.7
10	2020	19,4

The energy associated with solar constant per minute

$$= 1367\frac{W \times 60s}{m^2 \times 60s} = 1367 \times 60\frac{J}{m^2\, min}$$

$$= 1367 \times 60\frac{J}{m^2 \cdot min} \times \frac{1Cal}{4.187J} \times \frac{1m^2}{10^4 cm^2}$$

$$(\text{SeeAppendix1, 1cal} = 4.187\text{Jand1m}^2 = 104\text{cm}^2)$$

$$= 1.96\frac{cal}{min\, cm^2} \approx 2\frac{cal}{min\, cm^2}$$

The average solar energy in cal/min.cm^2 in extraterrestrial region $= 2\frac{cal}{min\, cm^2}$

Example 7.8 Evaluate an overall efficiency of photosynthesis occurring in the plant in light reaction during winter condition at noon time.

Solution

Since the maximum solar radiation in winter for North Indian conditions (terrestrial region) at noon time = 700 W/m^2

$$\text{So, an average solar radiation in winter} = 350\text{W}/\text{m}^2 \tag{7.8.1}$$

This is about 25% (0.25) = $\frac{350}{1367}$ of solar constant 1367 W/m^2

Hence solar radiation at noon time in winter in $\text{cal}/\text{min.cm}^2$in extra-terrestrial region

$$= 2\ \text{cal}/\text{min}/\text{cm}^2 \tag{7.8.2}$$

For the purposes of evaluating photosynthesis efficiency, it is necessary to daily calories per cm^2. There is 47–48% of solar energy in the visible range (Sect. 1.2) incident on the plant (see Fig. 1.6). The input solar energy available per day in visible wavelength (0.38 μm < λ < 0.78 μm) for photosynthesis for biomass production (biomass) is obtained as follows:

$$\text{Input solar energy per day in}(0.38\ \mu\text{m} < \lambda < 0.78\ \mu\text{m})\text{for 5 hours} =$$

$$0.5\frac{cal}{min \cdot cm^2} \times (0.47) \times \frac{60min}{1hr} \times \frac{5hr}{1day} \tag{7.8.3}$$

$$= 70.5\frac{cal}{cm^2\,day}(296\text{J}/\text{cm}^2\text{day})$$

For a typical winter day, the value of solar energy = 0.5 $cal/cm^2 \cdot min$. (Eq. 7.8.2)

The overall efficiency of photosynthesis is as follows:

(i) Plants use visible light (0.4–0.7 μm) = 47–48% which is about 50% of solar spectrum.

(ii) For three series absorption process of chloroplasts, each chloroplast absorbs 0.3 of incident radiation irrespective of any wavelength. The third chloroplast absorbs 59 W m^{-2} of radiation and the total absorbed radiation by the three chloroplasts is 263 W m^{-2} Chloroplasts absorption = $\frac{263}{350}$ = 75% (in fraction = 0.75) (Eq. 7.8.1)

(iii) Theoretical photon to glucose efficiency for absorption process; 30%

(iv) Dark respiration (metabolism) uses about 50%

By taking all of these factors into account,

$$\begin{aligned}\text{The overall light and dark efficiency} &= 0.5 \times 0.75 \times 0.30 \times 0.5\\ &= 0.056 \text{ in fraction}\\ &= 0.056 \times 100 = 5.6\%\end{aligned} \tag{7.8.4}$$

Example 7.9 Calculate the biomass produced per year for considering photon to glucose efficiency for absorption process; 30% and 9% land area of Earth's surface is covered by forest.

Solution

$$\text{Amount of solar radiation used for photosynthesis in cal/min/cm}^2$$
$$= 0.3 \times 70.5\ \text{cal}/(\text{cm}^2\text{day})$$
$$= 101\ \text{cal}/(\text{cm}^2\text{day})[\text{photon to glucose efficiency for absorption process; } 30\%,\ \text{given}][\text{Eq. 7.8.3}] \quad (7.9.1)$$

From Example 1.4 we have

$$\text{The surface area of the Earth} = 4(\pi) \times (6,500,000)^2 = 53,066 \times 10^{10} m^2$$
$$= 5.3 \times 10^{14}\text{m}^2$$

$$[\text{Radius of Earth} = \textit{about}\ 6500,000\text{m}$$

$$\text{There are 30\% area covered by land} = 0.3 \times 5.3 \times 10^{14} m^2 = 1.59 \times 10^{14} m^2$$

Further, 30% land should be covered by forestation, hence planted area on the Earth $= 0.30 \times 1.59 \times 10^{14} m^2 = 47.7 \times 10^{12} m^2 [0.09 \times 5.3 \times 10^{14} m^2 = 47.7 \times 10^{12} m^2]$

So, annual solar radiation used for photosynthesis on cultivated Earth $= 101 \frac{cal}{cm^2 day} \times 47.7 \times 10^{12} m^2 = 101 \times \frac{4.2J}{cm^2 day} \times 47.7 \times 10^{12} m^2 = 101 \times \frac{4.2 \times 10^4}{m^2 \times 5 \times 3600} \times 47.7 \times 10^{12} m^2 \times \frac{J}{s} = 11.24 \times 10^{15} \frac{J}{s}$

Here, it is important to note that five (5) hours per day has been considered.

If energy content of 3890 cal/gram (Table 7.6, cal/g = kcal/kg) with the dry biomass production resulting from photosynthesis is assumed then

$$\text{The biomass production per year} = 11.24 \times 10^{15} \frac{J}{s} \times \frac{1cal}{4.18J} \times \frac{6.57 \times 10^6 sec}{year} \times \frac{1gram}{3890cal}$$
$$= 45.42 \times \frac{10^{17} gram}{year} \approx 45.42 \times 10^{11} ty^{-1}$$

$$[\text{Five hours per day has been considered} \times 3600 \times 365 = 6.57 \times 10^6\text{s}]$$

Example 7.10a Calculate ash content in any biomass due to open combustion.

Solution

Ash content of any biomass can be evaluated by using the following formula

$$\text{Percentage}(\%)\ of\ ash\ content = \frac{Weight\ of\ ash}{weight\ of\ sample} \times 100 \qquad (7.10.1)$$

A small experiment was conducted by considering an amount of dry wooden stick of about 1.15 kg and burned in metal pot as shown in Fig. 7.9 to heat up the room in winter conditions and allowed to burn completely and then cooled down completely. We have done expedient for few more wooden materials.

The following observations have been made:

(i) Ash content in percentage depends on density of materials and its mass.
(ii) Biomass should be completely dried (no moisture content) to avoid carbon dioxide and carbon monoxide emission during burning as can be seen in Fig. 7.9.
(iii) The biomass should be cut into small pieces to ensure minimum moisture content.
(iv) Electronic balance should be used for an accurate value of biomass before and after burning.
(v) Ash contains maximum carbon and it is used as a fertilizer.

We weighed the ash content in an electronic scale and found it had about 233–240 gm variation. This may be due to variation in composition of biomass and some ash particulate came out of the pot with flame.

Fig. 7.9 Burning of biomass

$$\text{The percentage of ash content} = \frac{0.235}{1.15} \times 100 = 20\%$$

We found that the results are as per our expectation for dried wood stick. However, for dried grass, it is about 3–5%.

Example 7.10b Evaluate the mass dry wood and cow dung cake of with burner efficiency of 60% and 11% respectively for the temperature rise of water of about 40 °C of mass of 5 kg.

Solution

From Table 7.6, we have

Combustion (CO_V) value of dry wood ≈ 15,886 kJ/kg = 15,886 ×10^3 J/kg.

Combustion (CO_V) of cow dung cake = 8,753 kJ//kg= 8, 753 × 10^3 J/kg.

Burning efficiency (η_b) wood and cow dung cake are 60%and 11% respectively.

For dry wood, if M_{bio}inkg is mass of dry wood, then

M_w C_p $\Delta T = M_{bio} \times (CO_V) \times (\eta_b)$, Cp = 4200 J/kg °C (From approximate constant value).

or,

$$M_{bio} = \frac{M_w \mathrm{Cp} \Delta T}{(CO_V) \times (\eta_b)} = \frac{5 \times 4200 \times 40}{15,886 \times 10^3 \times 0.6} = 0.08kg = 80g$$

Similarly, for cow dung cake

$$M_{bio} = \frac{M_w \mathrm{Cp} \Delta T}{(CO_V) \times (\eta_b)} = \frac{5 \times 4200 \times 40}{8,753 \times 10^3 \times 0.11} = 0.872kg = 872g$$

So, one can observe that the cow dung cake requirement is ten times that of dry wood requirement due to its lower density.

With respect to the environment, following are advantages and disadvantages from the burning of biomass:

7.3.4.1 Advantages of Biomass

There are some of the advantages of biomass:

(a) **Renewability**: It is renewable source of energy because its production depends on solar radiation which is renewable. Biomass is always, widely and freely abundant throughout the world irrespective of developed and underdeveloped countries. It is a natural source of energy.

(b) **Carbon control**: It is carbon neutral biomass due to its absorption of CO_2 from environment for photosynthesis in the presence of light and rejection of CO_2 in the environment during the burning. It has carbon footprint of zero.

(c) **Reliability**: It reduces the dependency on fossil fuels which pollute. It is a reliable source of energy.

(d) **Economical**: It can be grown as a individual capacity and hence it is economical than fossil fuels and environmentally friendly. Further, it adds revenue to users for manufacturing material for construction of building (European countries having mostly cold climatic condition) countries which can be reused after its life ends. It has insulating properties and thermal loses through it is minimum.
(e) **Garbage in landfills**: It does not create due to its ashes being used as a fertilizer. There will be less waste in landfills.
(f) **Employment**: It generates employment of young people in rural areas from farming to forest management in desert land and unused land areas. The technologies for creating biomass energy are improving all the time to provide more employment. The employment is expected to increase for reasons such as (i) meeting of nation's energy demands by biomass resources, (ii) the labor forces requirement in cutting and harvesting agricultural as well as forest produces and (iii) labor requirements for transport of biomass resources and in the operation of conversion facilities.
(g) **Storage**: It can be referred to as storage of solar energy and can be easily transported in rural areas along highways. So, intermittent nature of solar energy can be mitigated by using biomass
(h) **Biomass boiler**: It reduces household CO_2 contributions by as much as *9.5 tons* each year to make it an incredibly green way to run homes using air thermal heating.

7.3.4.2 Disadvantages of Biomass [4]

There are also some disadvantages of biomass:

(a) **Burning**: During burning of biomass, it emits a lot of carbon dioxide (CO_2) along with nitrous oxide (N_2O) and carbon monoxide (CO). It has a negative impact on humans such as eye infections.
(b) **Health hazards**: Biomass production may result in major health concerns or hazards to humans due to a lack of understanding of the small particles generated during its manufacture.
(c) **Biomass production**: The full grown plants require a very large land area to operate and store. This means a higher land cost and therefore has greater limitations in urban areas.
(d) **Bio-fuel**: Sometimes, the production of bio-fuel alone **is not as efficient** as fossil fuels and is less economical due to requiring more biomass energy to burn the organic matter than it ends up producing. The labor requirement is 18 times more in million kilocuries of ethanol than an equivalent amount of gasoline. The possibilities of an increased employment are greater occupational hazards. Significantly, there are more occupational injuries and illnesses with biomass production in comparison to either/coal, oil or gas recovery operations.
(e) **Methane gas (CH_4)**: Methane gas (CH_4) contained in animal waste (cow/buffalo's dung) is around 30 times more responsible for global warming than carbon dioxide (CO_2). Burning natural materials like wood (see Fig. 7.9) also

emit nitrous oxide (N_2O) and carbon monoxide (CO) into the environment as a pollutants. Whilst biomass is carbon neutral it still release carbon dioxide (CO_2) into the air unlike wind and solar sources.

(f) **Deforestation**: Due to population increases since World War Two (see Fig. 6.5a, b) the demand for wood has also increased in domestic uses as well as in the manufacturing sector for construction of domestic tools and buildings in cold climates. This leads to deforestation and hence forest as a sink for carbon dioxide (CO_2) is decreasing daily. For increasing biomass production, various plants are planted and organized by governments and non government organization (NGOs), but trees grow monoculture crops. This growth strips nutrients from the soil and removes biodiversity. The algae formation is used to produce biomass, but this depends upon phosphorus fertilizers which damage water supplies for the surrounding areas.

(g) **Nutrient Removal and Losses**: Significant nutrient is lost during the harvesting of high energy value crop. There are 224 kg (N), 37 kg (P), 140 kg (K) and 6 kg (Ca) are the nutrients contained in both corn grain and residues for yield of 8000 kg/ha [5]. Hence, the mentioned nitrogen should be replaced for each subsequent crop. The amount of energy needed to replace the nutrients lost would be the equivalent to 46 liters of oil per hectare.

(h) **Loss of Natural Biota, Habitat and Wild Life**: The habitat and food sources of wildlife and other biota (living organism in a particular environment/climate) are changed due to conversion of the natural ecosystem into energy-crop plantations [3]. Cutting down of forests and wetlands for bio power generation reduces many habitats and mating areas of mammals, birds and other biota.

Monoculture plantations of fast-growing trees reduce (i) the diversity of vegetation and (ii) the areas as habitats for many wildlife species. These monocultures are less stable due to increased energy inputs in the form of pesticides and fertilizers to maintain productivity.

(i) **Social and Economic Impacts**: The major social impacts are (i) the shift in employment, (ii) increase in occupational health and (iii) safety problems. The requirements of direct labor inputs are 2–3 times greater per million kcal (1 kcal= 4.2 J) for wood biomass resources than fossil fuel coal. A wood-fired steam power plant requires 4 times as many construction workers as a coal-fired plant.

7.3.5 Geothermal Energy

Geothermal consists of geo (earth) and thermal (heat). Geothermal energy is **heat** available inside the Earth. It is a renewable resource source of energy due to large heat capacity like solar energy. It is clean, and environmentally and climate friendly like solar and wind. It is possible to use this energy for human uses. The hottest part of our planet is at 2900 km below: the Earth's crust. Thus, the availability of geothermal energy in terms of temperature can be classified as follows:

Class-I: High temperature geothermal energy: Temperature range (T) > 150 °C (used for electrical power production/heating).

Class-II: Average temperature geothermal energy: Temperature range (T) 90 °C < T < 150 °C (used for the electrical power production/heating).

Class-III: Low temperature geothermal energy: 30 °C < T < 90 °C (used for thermal heating).

Case-IV: Very low temperature geothermal energy: Temperature range (T) < 30 °C (used as the heat transfer fluid for heat pumps).

The advantages and disadvantages of geothermal energy [6] are discussed in the following section.

7.3.5.1 Advantages

The following advantages can be observed:

(a) **Availability**: It is always available, unlike solar and wind. *Geothermal power plants* produces electrical energy for around 8,600 hours per year if available on a given site. On the other hand, solar energy is available around only 2,000 hours per year (5.4 hours per day). The planning to install geothermal is more accurate than other renewable energy sources. It is largely untapped and sustainable. It can be directly used for domestic purposes (space heating [7], greenhouse heating and irrigation in cold countries, bathing, warmth and cooking etc.).

(b) **Space**: It does not require a large space because it is stored inside the Earth, unlike solar and wind energy. There is a modest space requirement for thermal energy through the Earth Air Heat Exchanger (EAHE) buried underground, and the heat pump. Very little space is needed above ground for the cooling tower and a steam turbine.

(c) **Silent energy**: The noise can be only heard during construction. After installation there is no noise involved during operation and maintenance, hence it is silent unlike hydropower plants. This applies to both domestic as well as large scale use

(d) **Job opportunities**: It can create 34 jobs per installed megawatt, in comparison to other renewable energy plants, which generate only 19 jobs per megawatt for wind power, and 12 for photo voltaic systems, in terms of operation and maintenance.

(e) **Double recycling**: As mentioned earlier, the embodied energy in any system plays an important role at end of life of any system. In this case, the materials used can be recycled after dismantling the system at end of life. Further, the discharged hot water available from the power plant system can be reused either in the same plant or can be directly supplied for domestic applications.

(f) **Maintenance**: For domestic applications such as hot water, the geothermal unit is a closed loop requiring little maintenance due to the pressure of self-regulation of the fluid in the piping.

(g) **Safety and reliability**: It is safe, reliable and has longevity due to its lifespan of up to 100 years, which is life of most of RRC buildings. It becomes economical if integrated to buildings for thermal heating and cooling, compared to solar thermal and PV systems. There is no risk of fire, hence life is safe. For example, underground water temperatures at about 100 m for northern Indian condition is around 25 °C, and boring supplies water at 25 °C all through the year. In summer, the water seems to be cool against 45 °C ambient air temperature and hot against 10–15 °C in the winter. In Turkey, hot springs are available everywhere and hence each house has heating from hot springs.

(h) **Renewability and potential**: It is a source of renewable energy. It will be available as long as the Earth exits. Presently, the global energy demand is around 15 terawatts, and the total potential energy available from geothermal sources is much more.

7.3.5.2 Disadvantages

The following disadvantages can be observed:

(a) **Environment (air pollution)**: The geothermal water/steam contains greenhouse gases (GHG), namely hydrogen sulfide (H_2S), carbon dioxide (CO_2), methane (CH_4) and ammonia (NH3) and hydrogen (H_2). The extraction of geothermal energy from the ground leads to releases of these greenhouse gases, polluting the environment surrounding the geothermal plant. But, the amount of greenhouse gases (GHG) released is **much lower than** in the case **of fossil fuels**. It ranges from 1.0 to 50 g/kg of steam. Hence, one can say it is environmentally friendly.

(b) **Location**: Geothermal energy is not equally distributed all over the world. It can be harnessed only at locations of its availability, such as hot springs (only 15 locations in the world) which are available in Turkey, India, Iceland, Alaska, Wyoming National Park, Antarctica, China, Indonesia, Italy, Greenland, Japan, California, New Zealand, Chile, and Colorado.

(c) **High costs**: Geothermal energy for electrical power plant is expensive to tap into in comparison to thermal energy for domestic applications. It costs around $2–$7 million per megawatt capacity which is difficult for developing countries to fund.

(d) **Water pollution**: Water pollution in power production is a potential hazard for rivers and lakes in vapor-dominated reservoirs. Waste steam constitutes 20% of the steam supply, and this must be added to the waste water. It carries a variety of toxic chemicals [arsenic (As), mercury (Hg), lead (Pb), zinc (Zn), boron (B) and sulphur (S), together with large amounts of carbonates [**CO3(2−)**], silica (SiO_2), sulphates [SO_4^2-] and chlorides (Cl−). The pollution of water bodies is

more easily controlled than in water-dominated reservoirs. It may contain large quantities of salts (NaCl above 300 g/kg to 70 kg/kWh).

(e) **Land subsidence**: The weight of the rocks above a reservoir of groundwater, oil or geothermal fluids is borne by the mineral skeleton of the reservoir rock and by fluids in the rock pores. As fluids are removed, the pore pressure is reduced and the ground then tends to subside. Less subsidence is expected with harder reservoir rock. It can damage roads, buildings, pipelines and drainage systems etc. The method of drilling geothermal plant deep under Earth is dangerous to the people working surrounding the plant during installation.

(f) **Induced Seismicity**: Many geothermal reservoirs are located in geologically unstable zones of the Earth's crust at high temperatures which are characterized by volcanic activity, deep earthquakes, seismic activity, tectonic stress and heat flow.

(g) **Noise**: A noise level of 90–122 dB at free discharge and 75–90 dB through silencers during maintenance of exhausts, blow downs and centrifugal separation are observed, all of which are health hazards. These can be avoided by the installation of silencers on some equipment.

(h) **Fog**: Large volumes of steam escaping into the air during operation of the plant can cause dense fog to occur. It may drift across nearby roads causing traffic hazards.

7.3.6 Ocean Thermal Energy Conversion (OTEC)

Ocean Thermal Energy Conversion (OTEC) systems depend *ocean thermal gradient* between cooler regions in deeper areas of sea, and warmer regions at shallower depths of sea to operate a *heat engine* and produce electricity. There are two types of OTEC systems: **(a) closed system**: The *ammonia* or *R-134a* uses fluids as *refrigerants* with low boiling points to generate electricity and (b) **Open system**: In this case, the working fluid of the engine is *seawater* itself with a very high *capacity factor* to operate in *base load* mode. At present, the most favorable used heat cycle for OTEC is the *Rankine cycle* using a low-pressure turbine under a closed system. The ideal thermal efficiency of OTEC is between 50 and 60%, but viable OTEC system efficiency is characterized by Carnot efficiencies in the range of 6–8%, due to a low operating temperature range.

Following are advantages and disadvantages of ocean thermal energy conversion (OTEC) plants.

7.3.6.1 Advantages of Ocean Thermal Energy (OTE)

(a) **Renewable Energy**: It is a green (clean) and renewable energy source like solar energy. It is accessible in only a few countries worldwide. The variance in temperatures between hot and cold will always exist. So, one can generate electricity indefinitely without this source draining. This temperature difference depend on hydrolic cycle, Fig. 2.1 due to solar radiation.
(b) **Clean energy**: As per the Kyoto protocol and the Paris agreement in 2015, it has been decided to use green/clean energy, combining plantations and electrical power production through renewable energy sources. Clean energy production is an aim of the future for various countries. Ocean thermal energy (OTE) is one such clean energy source.
(c) **Reliable and environmental friendly**: The ocean thermal energy conversion (OTEC) plant is a reliable source of energy due to having minimal variations, unlike solar and wind. It can be used to harness electrical power continuously. The OTEC plants are mainly situated in sea areas away from cities and hence there is no interaction between the plants the dense human populations of such cities. The installation of OTEC plants doesn't need fossil fuels.
(d) **Low Maintenance**: The machinery utilized for installation of ocean thermal energy conversion (OTEC) needs low maintenance. The floating OTEC plants are primarily minor and hence cheap.
(e) **Independent of Weather**: Only the top portion of OTEC appears floating on the sea surface. Therefore, the weather doesn't significantly affect these plants.

7.3.6.2 Disadvantages of Ocean Thermal Energy

(a) **High Cost and site of Production**: The cost of installation is very high due to specific locations for installation of OTEC. Establishing ocean thermal energy conversion (OTEC) is limited to some developed countries due to the requirement of skilled professionals to install and maintain the plants. It affects the ocean water quality.
(b) **Interfere with navigation**: The ocean thermal energy conversion (OTEC) plant floats on the water surface and can disturb navigation nearby.
(c) **Slight temperature difference**: The discrepancy in temperatures between surface water and deep water might be small. So, this may generate electric energy with moderate efficiency (6–8%). It can also be very costly to small electricity businesses.
(d) **Large size turbines with expensive liquid**: The ocean thermal energy conversion (OTEC) plant might need large turbines. These plants employ massive turbines due to the low stress in the boiling propane.

(e) **Harmful on Marine life**: The ocean thermal energy plant has pipes running down to the deepest parts of the ocean. Thus we can see that these pipes might disturb marine life. Also, small water animals might get drawn in by the force in the pipes through the pumping action of the water. The discharge from OTEC plants affects the (i) surrounding marine environment due to heating of water, (ii) the release of toxic chemicals, (iii) impingement of organisms on intake screens and (iv) entrainment of small organisms by intake pipes etc. Further, a large discharge of water (mixed warm and cold water) near the surface creates a plume of sinking cool water up to 3–4 °C. This causes high mortality among corals and fishes. Toxic chemicals (ammonia and chlorine) from OTEC plants may enter the environment, killing local marine organisms.

7.3.7 *Tidal Energy*

Tidal energy is a form of power produced by the natural rise and fall of tides (periodic variation) caused by the gravitational force/attraction exerted by celestial bodies, namely the planet Earth, the Sun, and the Moon. Tide energy is also classified as renewable energy due to gravitational interaction with the Moon and Sun and the Earth's rotation, which are inexhaustible. In this case, kinetic energy the tide is due to vertically moving water converted into mechanical power (turbine) and then electrical power by generator, just like in the case of wind energy/power. Tidal barrages/dams are constructed with a narrow opening to the sea water. With sea water level rises, the water rushes into the tidal barrages/dams to move the blades of the turbines which are attached at the openings of the dam for mechanical power. Thus, the mechanical power is converted into electricity by the generation. Tidal barrages rely on the flow of water in and out of estuaries to produce energy.

There are some advantages and disadvantages of tidal energy (TE).

7.3.7.1 Advantages

These are as follows.

(a) **Environment-friendly**: It is an environmentally friendly and clean source of energy due to the potential energy difference in consecutive tides. It does not produce any harmful gas and occupies a very small space for energy production. Vertical-axis turbines and coastal turbines are more affordable for lower environmental impacts. There is a reduction in greenhouse gas (GHG) emission like solar and wind.

(b) **Predictable energy source**: Tidal currents/waves are highly predictable due to high and low tides in the ocean as per periodic cycles due to the motion of the Sun. Power can be generated even at speed of 1 m/s due to its high density of water, unlike air. Advance tidal turbines have 80% efficiency.

(c) **Operational and maintenance**; It is cost effective: Drum fire reduces the damage of high tidal surge on the ground.
(d) **Life of equipment**: The life span of tidal power plants is as much as 100 years due to concrete structures spread across river estuaries. For example, La Rance plant in France was installed in the year 1966, and continues to create clean energy now.

7.3.7.2 Disadvantages

Some of the disadvantages of tidal energy are as follows:

(a) **Construction costs**: The tidal energy (TE) power plant is still a relatively new technology. The growth of TE power plants has not been on the same pattern as other sources of energy technology namely wind/geothermal/solar etc. So, the TE technology is comparatively expensive. **The electricity generation by TE technology is more costly than other renewable technology**. This can be economical if systems are installed on a large scale by using more advanced technology.
(b) **Location limits**: The tidal energy (TE) systems have to be installed close to the land along our coastlines due to strongest currents. As it stands, it is difficult and more costly to transport energy longer distances. This is because many fast tidal flows are in areas near shipping lanes and, in some cases, too far from the grid. This is another factor that holds this form of energy back. However, there is hope that technology will evolve so that tidal energy systems can be located out at sea. On the other hand, *unlike hydropower*, tidal does not result in the flooding of land mass.
(c) **Marine life**: Construction of tidal barrage, dams, and turbines and generators and other systems can harm marine and sea life. However, the barrages alter water flow through the tidal turbines. This turbine disrupts the way marine life thrives and navigates low and high tides. The impacts of tidal plants on the environment include ports and navigation, facilities for recreation, water quality, agriculture and land drainage, visual amenity, bird life, migratory fish, the ecosystem in the estuary etc. It is also feared that tidal power plants hamper the other natural uses of estuaries such as fishing or navigation.

7.3.8 Oceanic Wave Energy

It is another form of renewable energy. It is harnessed from the motion of the waves in the ocean. It is also referred to as ocean **wave energy and** carries a huge amounts of power. Basically, it is an indirect source of solar energy which creates wind air (Sect. 2.7, Fig. 2.6) and the action of winds blowing over the ocean's surface is partially responsible for wave energy. Another reason for wave energy is deep-water waves moving towards shallow water, that change into breaking waves. The amount of power (P) from wave energy depends on (i) wavelength height (amplitude)/wave front in m (H), (ii) density of sea water (ρ) in kg/m^3 (iii) time period (frequency) in s$^-$, and (iv) accelerations due to gravity, g = 9.8 m/s^2.

An expression for power derived at deep-sea wave water per unit wave front length is given [8] by

$$P = \frac{1}{64\pi}(\rho g^2)H^2T \approx \left[500\frac{W}{(m^3 \cdot s)}\right]H^2T\left(\frac{W}{(m^3 \cdot s)}\right)$$
$$\approx \left[0.5\frac{kW}{(m^3 \cdot s)}\right]H^2T(in\frac{kW}{m}) \tag{7.3a}$$

Wave power for wave front length L is given by

$$P = \frac{1}{64\pi}(\rho g^2)H^2TL \approx \left[500\frac{W}{(m^3 \cdot s)}\right]H^2TL \approx \left[0.5\frac{kW}{(m^3 \cdot s)}\right]H^2TL \tag{7.3b}$$

The mean wave energy density per unit horizontal area (E) in J/m^2 which is sum of kinetic and potential energy density per unit horizontal area is given by

$$E = \frac{1}{16\pi}(\rho g)H^2 \tag{7.4}$$

From above one can observe that the average energy density per unit area of sea surface waves is proportional to the square of wave height (H). The potential energy density is equal to the kinetic energy, both contributing half to the wave energy density E.

For equal wavelengths, a wave with larger amplitude gives more energy when it falls back to sea level in comparison to a lower amplitude. One can observe that if height is increased to double then produced energy is increased by four times.

7.3.8.1 Advantages of Oceanic Wave Energy

(a) **Renewable**: It will never run out as long as the Sun and ocean exist. It is always available, and is clean and green. Coastal well-populated regions can harness energy from wave energy plants.
(b) **Environment friendly**: It creates mechanical and electrical power from oceanic waves with no harmful byproducts like GHG gases, waste, and air pollution.
(c) **Predictable**: It is consistent, reliable and proves to better than other sources of renewable like wind or Sun.
(d) **Land**: There is a requirement of land for system installation.

7.3.8.2 Disadvantages of Oceanic Wave Energy

(a) **Suitable to Certain Locations**: It is only suitable in populated cities (town) near to coastal regions.
(b) **Effect on Marine Ecosystem**: Installed turbine and generator machines disturb the seafloor, change the habitat of near-shore affecting crabs and starfish, creates noise which disturbs the sea life around them and there is also a danger of used toxic chemicals on wave energy platforms spilling and *polluting the water* sea water nearby. Large scale plants needs very large areas of the surface of the ocean with wave energy devices, potentially harming marine life.
(c) **Wavelength**: One can see that wind power is highly dependent on wavelength (Eq. 7.3) which requires a consistent flow of powerful waves to generate electrical power. It becomes unpredictable to forecast accurately in some areas and then becomes unreliable.
(d) **High Costs**: It needs the enormous cost of electrical power production due to considerable set-up with uncertain lifespan. Strong waves can damage equipment with little chance of repair.
(e) **Noise**: Wave energy conversion systems make noise. The levels are expected to be below the levels of a normal ship. They are expected to be noisier than the surrounding wind and waves with full operation of plant. Any noise generated can travel long distances underwater which can have an impact on certain animals such as seals or whales.
(f) **Threat to Navigation**: The wave energy conversion devices may be a dangerous obstacle to any navigational craft. They cannot be seen or detected by radar or direct sighting. The use of adequate navigational aids on offshore wave energy conversion devices is required.
(g) **Coastal erosion**: Some energy conversion devices concentrate wave energy into a tapered area. This concentrating surge device channels waves to increase wave height for redirection into elevated reservoirs. The water from elevated reservoir then passes through hydroelectric turbines on the way back to sea level, generating electricity. Continuous arrays of such onshore or shore-based wave-energy devices physically alters coastlines.

Example 7.11 Determine power per unit length in with a wave height of 3 m and a wave energy period of 8 s.

Solution

From Eq. (7.3a), we have an expression for power per length as

$$P = \left[0.5\frac{kW}{\left(m^3.s\right)}\right]H^2T(in\frac{kW}{m}) = 0.5 \times 9 \times 8 = 36kW\text{perm}$$

Example 7.12 Determine the length of wave power of 30 kW crossing the great height of 20 m in 1 s.

Solution

Known parameters:

P = 30 W, H = 20 m and Wave period T = 1 s.

The Wave power formula from Eq. (7.3b) is given by,

$$P = \frac{1}{64\pi}(\rho g^2)H^2TL$$

or,

$$L = \frac{P}{0.5 \times H^2T} = \frac{30}{0.5 \times (20)^2 \times 1} = 0.15cm = 0.15m \qquad \text{[Eq. 7.3 b]}$$

$$= (64 \times 3.14 \times 30)/(1.025 \times 1 \times 9.8^2 \times 20^2) = 0.15m \qquad \text{[Eq. 7.3 b]}$$

7.3.9 *Biogas Plant*

Biogas consists of various gases namely *methane* (CH_4, 59%), *carbon dioxide* (CO_2, 40%) and *hydrogen sulphide* (H_2S, 0.08%) and remaining water vapor. Biogas when combusted in the presence of oxygen (O_2), it produces heat energy thet can be used for many applications like cooking, lighting and electrical power generation. It is produced from animal (cow, buffalo etc.) *manure*, fresh *agricultural/vegetables waste* materials, *municipal waste*, fresh *plant materials*, *sewage*, *green waste* and *food waste* etc. by using biogas plants all input materials to biogas plant as mentioned earlier is directly and indirectly produced by solar energy and hence it is a *renewable energy* source, and biogas can be treated indirectly as storage of solar energy. Here, we should clearly note that:

(a) The input materials for biogas plant (see Fig. 7.10) as a feed is produced directly in the presence of solar radiation.
(b) Biogas in biogas plant is produced in the absence of oxygen at optimum temperature of 37°C

There are two type of biogas plant: floating and fixed dome (see Fig. 7.10). For more details, consult a review paper on biogas plants [9].

7.3.9.1 Advantages of Biogas

(a) **Eco-Friendly**: It is a renewable, green, clean energy and friendly with environments and climate. Gas produced through bio-digestion in the absence of oxygen is non-polluting, i.e. zero emission of GHG. It combats global warming. The

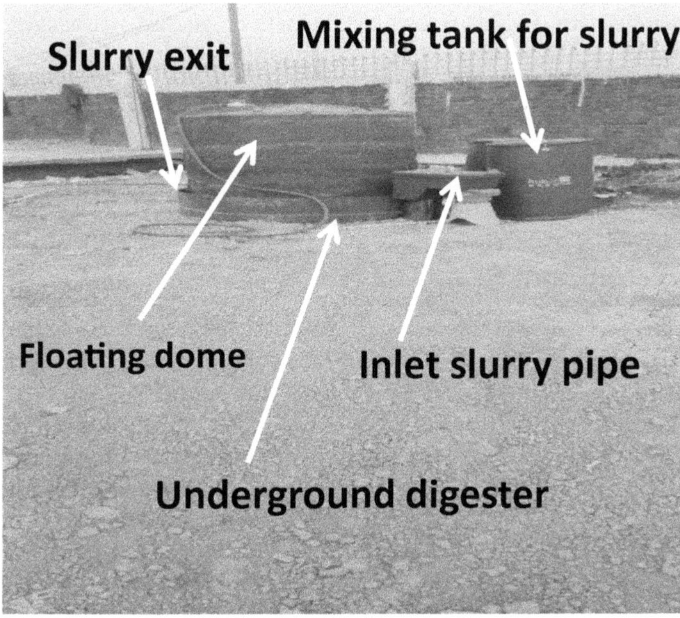

Fig. 7.10 Floating biogas plant at Ballia (UP) India

use of biogas reduces the burden of fossil fuels. No additional input required except the operating temperature of 37 °C.

(b) **Reduction of soil and water pollution**: Landfills consists of waste material collected from various corners of cities in different forms and spread foul smells along with allowing toxic liquids to pass on into underground water sources. But in the case of biogas plants, there is no chance of underground water pollution due to the digester buried underground as shown in Fig. 7.10. An anaerobic digestion (in absence of oxygen) deactivates pathogens and parasites which effectively reduces the incidence of waterborne diseases.

(c) **Organic fertilizer**: The by products of biogas generation exit through the outlet of the plant as shown in Fig. 7.10, and it is an enriched organic compound which can be used for many purposes such as chemical fertilizer and aquatic feed (fish). The fertilizer discharge from the exit of the digester can accelerate plant growth and resilience to diseases.

(d) **Low-cost technology**: It is a very simple and economical plant and it can be designed and installed by local technicians with low embodied energy like other renewable energy plants. It needs little investment at small scale except for feed to the biogas plant. Small biodigesters made of FRP (fiber re-enforced plastic material) can be used at home, utilizing kitchen waste and animal manure**.** In large scale biogas plants, biogas can also be compressed to achieve the quality of natural gas. It becomes easy to transport larger distances like LPG gas cylinders.

(e) **Healthy cooking**: Biogas from biogas plants is quickly combustible to avoid polluted air by using biomass for cooking purposes. This also saves collecting firewood.

7.3.9.2 Disadvantages

(a) **Efficiency**: The biogas plant is not efficient due to its temperature dependence of anaerobic digestion process. It does not work in cold climatic conditions.
(b) **Input**: It is very difficult to have large scale biogas plants due to non-availability of sufficient slurry for input materials. It can be constructed only after surveys regarding any type of slurry material to avoid continuous operation. Due to impurities present in biogas, it cannot be used in automobiles due to corrosion.
(c) **Temperature**: As mentioned above, the optimum temperature of a biogas plant needs temperature of 37 **°C** for activation of bacteria. So, it cannot operate in cold climatic conditions. In this case, some active heating arrangement has to be made for slurry in the digester, as suggested by Tiwari et al. [9].
(d) **Dense Metropolitan Areas**: Due to non-availability of sufficient raw materials such as food waste, manure etc. biogas plants are not suitable in cities, but biogas generation is suitable for rural and suburban areas.

Table 7.8 gives the proportions and multiplication factors for different biogas plant to get appropriate size of digester [10].

Example 7.13 Calculate the volume of digester of fixed dome type biogas plant to produce 3 m^3 biogas.

Solution

From Table 7.8, we have

Multiplication factors to convert gas production to plant volume for fixed dome plant = 3.

Hence total fixed dome plant volume $= 3 \times 3 = 9\ m^3$

Example 7.14 Calculate the volume of digester of floating dome type biogas plant, Fig. 7.10 to produce 3 m^3 biogas.

Table 7.8 Proportions and multiplication factors to convert total plant volume and rated daily gas production into digester volume and gas storage volume (Digester volume (DV); Gas storage volume (GSV); Total volume (TV))

Biogas plant type	Fixed dome plant DV GSV TV	Floating drum plant DV GSV TV	Balloon/bag digester DV GSV TV
Digester and gas storage volumes as a share of total plant volume	80% 20% 100%	70% 30% 100%	75% 25% 100%
Multiplication factors to convert gas production to plant volume	2.4 0.6 3.0	1.4 0.6 2.0	1.8 0.6 2.4

Solution

From Table 7.8, we have

Multiplication factors to convert gas production to plant volume for fixed dome plant = 2.

Hence total fixed dome plant volume = $3 \times 2 = 6\ m^3$

From Examples 7.13 and 7.14 one can observe that digester volume of floating dome type biogas plants are more cost effective than fixed dome type biogas plant for same volume of biogas production. This indicates that floating dome type biogas plant is cost effective than fixed dome type biogas plant for same volume of biogas production.

Example 7.15 Evaluate the power available from biogas plant for following given parameters:

Number of animals = 5; Dry matter consumed per day per animal = 2 kg/day; Number of animal = 5; Efficiency of biogas burner = 60% (0.60 in fraction), Table 7.6; Heat combustion of methane = 28 MJ/m^3; Methane fraction in biogas = 0.6 and biogas production per dry mass = 0.24 m^3/kg.

Solution

We know that

$$\text{Total dry mass consumed by animals} = \text{Dry matter consumed number of animals} \tag{7.15.1}$$

and,

The power available from biogas digester = Efficiency of biogas burner

$$\begin{aligned}(\eta_b) \times &\ \text{H heat combustion of methane}\left(\text{MJ}/\text{m}^3\right)\\ &\times \text{Methane fraction in biogas combustion of methane}\left(\text{MJ}/\text{m}^3\right)\\ &\times \text{Methane fraction in biogas}\end{aligned} \tag{7.15.2}$$

From Eq. (7.15.1), one can have

Total dry mass consumed per day by 5 animals = 2 kg/day × 5 animals = 10 kg/day.

Biogas volume produced by biogas plant = 0.24 m^3/kg × 10 kg/day = 2.4 m^3/day.

Now from Eq. (7.5.2), we get after substituting of given vlues of parameters.

Power available from biogas digester = $0.60 \times 28\text{MJ}/\text{m}^3 \times 0.6 \times 2.4 = 24.192$ MJ per day = 24.192 × 0.27778 kWh = 6.72 kWh.

7.4 Conclusions and Recommendations

The following conclusions and recommendations have been made:

(a) The renewable energy available at any site should ideally be used in nearby villages and cities to avoid transmission and distribution losses.
(b) The hydro and biomass energy can be used continuously unlike solar and wind; these are indirect storage of solar energy. The big hydropower plant should be directly linked to grid power unlike biomass, biogas and geothermal energy sources.
(c) OTEC tidal and wave energy production is only site specification in coastal regions. It can be economical and should not be used for long distances due to transmission and distribution losses.
(d) Biogas plant is most successful for domestic applications as well as decentralized level with efficient input parameters like slurry. Biogas can also be referred to as indirect storage of solar energy.
(e) Awareness programs should be carried out at national and international levels among the users of renewable energy sources.
(f) Fossil fuels should be conserved as far as possible according to the Kyoto protocol and the Paris agreement on priority bases to avoid energy crises as well as natural disasters.
(g) Nature helps to protect wildlife including humans on the planet Earth.

Problems

Problem 7.1 Evaluate carbon di-oxide (CO_2) concentration in micro-gram per m^3 ($\mu g/m^3$) for a given 320 ppm of CO_2 in air at 25 °C and 1 atmospheric pressure.

Hint: Example 7.2 and Table 7.1b.

Problem 7.2 Evaluate ozone (O_3) concentration in micro-gram per m^3 ($\mu g/m^3$) for a given 20 ppm of ozone (O_3) in air at 25 °C and 1 atmospheric pressure.

Hint: Example 7.2 and Table 7.1b.

Problem 7.3 Evaluate the rate of evaporative in W/m^2 and amount of water evaporated in m^3 and heat transfer coefficient in W/m^2 °C from water surface (35°C) to an ambient air temperature (25°C) with a relative humidity of 50%. And compare the results with Example 7.3.

Hint: See Example 7.3.

Problem 7.4 Plot the curve between water mass evaporated with different value of relative humidity (0.25, 0.50, 0.79 and i).

Hint: See Problem 7.3.

Problem 7.5 Evaluate the amount of water needed in m^3 to compensate for seepage losses through different materials for Table 7.5 surface during 12 months for water pond surface area of 2,000 m^2 and compare the results for different materials.

Solution: See Example 7.6.

Problem 7.6 Find the biomass produced per year for considering photon to glucose efficiency for absorption process; 30% and 12% land area of Earth surface is utilized for forestation.

Hint: See Example 7.9.

Problem 7.7 Determine power in kW with a wave height of 6 m and a wave energy period of 810 s for wave front length of 0.2 m.

Hint: See Example 7.11 and Eq. (7.3b).

Problem 7.8 Determine power per unit length in W with a wave height of 3 m and a wave energy period of 8 s.

Hint: See Example 7.11 and Eq. (7.3b)

Problem 7.9 The mean wave energy density per unit horizontal area (E) in J/m^2 for Problems 7.7 and 7.8 respectively.

Hint: Use Eq. (7.4).

Problem 7.10 Problem 7.10 Repeat Example 7.10b for other biomass fuel of Table 7.7 and compare the results.

Hint: See Example 7.7.

Problem 7.11 Calculate the volume of digester of balloon/bag digester of biogas plant to produce 3 m^3 biogas.

Hint: See Example 7.11 and use Table 7.8.

Problem 7.12 Plot the curve between power produced per day with a fraction of methane gas in biogas for data given in Example 7.5.

Hint: Use Example 7.15.

Problem 7.13 Plot the curve between powers produced per day with number of animals for a given fraction of methane (0.60) in biogas for data's given in Example 7.5

Hint: Use Example 7.15

Objective Questions

7.1 Environment is also known as

(a) Atmosphere; (b) Biosphere

(c) Troposphere; (d) All of these

Answer: (b).

7.2 The hydrological cycle (H_2O) is

(a) Directly related to carbon cycle (C); (b) Directly related to nitrogen cycle (N)

(c) Directly related to sulfur cycle (S); (d) Indirectly related to carbon cycle (C)

Answer: (a) (see Chap. 2).

7.3 The following gas has maximum concentration in atmosphere

(a) Carbon dioxide (CO_2); (b) Oxygen (O_2)

(c) Nitrogen (N_2); (d) Sulfur dioxide (SO_2)

Answer: (c).

7.4 The following gases have the minimum concentration in atmosphere

(a) Carbon dioxide (CO_2);(b) Oxygen (O_2)

(c) Nitric oxide (NO); (d) Sulfur dioxide (SO_2)

Answer: (c).

7.5 The biosphere exists due to presence of

(a) Wind energy; (b) Solar energy

(c) Tidal and wave; (d) None

Answer: (c).

7.6 Reason for acid rain is the pressure of following gas in atmosphere

(a) Methane (CH_4); (b) Hydrogen (H_2)

(c) Amonia (NH_3); (d) Hydrogen sulfide (H_2S)

Answer: (b).

7.7 The non-condensable gases in a stream of geothermal field is

(a) Methane (CH_4); (b) Hydrogen (H_2)

(c) Amonia (NH_3); (d) Hydrogen sulfide (H_2S)

Answer: (a).

7.8 The carbon dioxide (CO_2) gas in a stream of a geothermal field is a

(a) Major contributor for polluting environment; (b) Minor contributor for polluting environment

(c) No effect for polluting environment; (d) A non-condensable gases

Answer: (a).

7.9 The Ocean Thermal Energy Conversion OTEC is more responsible for polluting

(a) Biosphere ecosystem; (b) Ocean ecosystem

(c) Both (a) and (b); (d) None of these

Answer: (b).

7.10 The tidal power (TP) plant hampers

(a) Fishing; (b) Navigation

(c) Both (a) and (b); (d) None of these

Answer: (c).

7.11 The wave energy power per unit length is

(a) Proportional to square of height; (b) Inversely proportional to height

(c) Equal to height; (d) All of these

Answer: (a).

7.12 The wave energy power per unit length is

(a) Proportional to square of height; (b) Proportional to wave energy period

(c) Equal to height; (d) All of these

Answer: (a) and (b).

7.13 The mean wave energy density per unit horizontal area (E) in J/m^2 is

(a) Proportional to square of height; (b) Proportional to wave energy period

(c) Equal to height; (d) All of these

Answer: (a).

7.14 The OTEC, tidal and wave energy is available

(a) Through out; (b) In sea region

(c) In oceanic region: (d) All over world

Answer: (b) and (c).

7.15 The OTEC, tidal and wave energy is environmentally friendly

Like solar energy; (b) Like wind energy

(c) Like biomass energy; (d) None of these

Answer: (d).

References

1. D.L. Hartley, Perspectives on renewable energy and the environment, in *Energy and the environment*, ed. by J.W. Tester, W. Do, N.A. Ferrari (MIT, Massachusetts, 1990)
2. I. Dincer, Renewable energy, environment and sustainable development, in *Proceedings of the world renewable Energy Congress, 20–55 September, Florence, Italy* (1998), pp. 2559–62
3. M. Gadgil, Biodiversity and India's degraded lands. Ambio **22**(2–3), 167–172 (1993)
4. H.L.S. Tandon, Recycling of crop, animal and industrial wastes in agriculture, fertilizer development and consultation organization (New Delhi, 1995)
5. D. Pimental, Frqiend C. Olson L, S. Schmidt, W.B. Wagner, K. Johnson, A. Westman, A.M. Whelan, K. Fegalia, P. Poole, T. Klein, R. Sobin, A. Bochner, Biomass energy: environmental and social costs. Environmental Biology Report, Cornell University Ithaca (New York, 1983), pp. 2–83
6. D.W. Layton, L.R. Anspaugh, K.D. O'Banion, Health and environmental effects, documents on geothermal energy—1981, Lawrence Livemore Laboratory, California, Report No. UCRL-53232 (1981), p. 61
7. Gheorghe Ilisei, Modeling an energy efficient house—DiVA portal diva-portal.org.·https://www.diva-portal.org
8. G.P. Thomas, Wave power—the theory behind ocean waves. https://www.azocleantech.com
9. G.N.Tiwai, R.K. Mishra, A.K. Singh, Thermal modeling of solar heating of bio-gas plant: a review. Int. J. Amb. Energy (2022). Taylor and Francis
10. IRENA, Measuring small-scale biogas capacity and production, International Renewable Energy Agency (IRENA), Abu Dhabi (2016)

Additional References

11. Environmental Impacts of Hydropower and Alternative. http://www.lupinepublishers.com
12. Dealing with Sediment: Effects on Dams and Hydropower. http://www.hydroreview.com
13. Wind Energy: The Pros and Cons. https://justenergy.com
14. Biomass—Wikipedia. https://en.wikipedia.org
15. Basal metabolic rate—Wikipedia. http://www.wikipedia.org
16. All the advantages of geothermal energy—Enel Green Power. https://www.enelgreenpower.com
17. Hot Springs Around the World That Are All About the. https://www.cntraveler.com
18. What are the Advantages and Disadvantages of Geothermal. http://www.twi-global.com
19. Wave Power—Vedantu. https://www.vedantu.com

Appendix

Conversion of Units

(a) Length, m
1 yd (yard) = 3 ft = 36 in (inches) = 0.9144 m
1 m = 39.3701 in = 3.280839 ft = 1.093613 yd = 1650763.73 wavelength
1 ft =12 in = 0.3048 m
1 in = 2.54 cm = 25.4 mm
1 mil = 2.54 10^{-3} cm
1 μm = 10^{-6} m
1 nm = 10^{-9} m= 10^{-3} μm
(b) Area, m^2
1 ft^2 = 0.0929 m^2
1 in^2 = 6.452 cm^2 = 0.00064516 m^2
1 cm^2 = 10^{-4} m^2 = 10.764 10^{-4} ft^2 = 0.1550 in^2
1 ha = 10,000 m^2
100 m^2 = 1 area, 10000 m^2 = 100 areas = 1 hectare (ha)
(c) Volume, m^3
1 ft^3 = 0.02832 m^3 = 28.3168 l (litre)
1 in^3 = 16.39 cm^3 = 1.639 10^2 l
1 yd^3 = 0.764555 m^3 = 7.646 10^2 l
1 UK gallon = 4.54609 l

(continued)

G. N. Tiwari, *Energy, Ecology and Environment*,
https://doi.org/10.1007/978-981-99-3997-8

(continued)

1 US gallon = 3.785 l = 0.1337 ft^3
1 m^3 = 1.000 106 cm^3 = 2.642 10$^{1\ 2}$ US gallons = 109 l
1 l = 10^{-3} m^3
1 fluid ounce = 28.41 cm^3
1 Cub feet =0.0283 m^3
(d) Mass, kg
1 kg = 2.20462 lb = 0.068522 slug
1 ton (long) = 1016.05 kg
1 lb = 16 oz (ounces) = 0.4536 kg
1 oz = 28.3495 g
1 quintal = 100 kg
1 kg = 1000 g =10,000 mg
1 μg = 10^{-6} g
1 ng = 10^{-9} g
One short tons = 0.9071847 metric tons
1 metric ton = 1000kg
One scf (standard cubic feet)=one cub feet =28.32 kg
(e) Density, specific volumes and energy density kg/m^3, m^3/kg, kWh/kg
1 lb/ft^3 = 16.0185 kg/m^3 = 5.787 10^{-4} lb/in^3
1 g/cm^3 = 10^3 kg/m^3 = 62.43 lb/ft^3
1 lb/ft^3 = 0.016 g/cm^3 = 16 kg/m^3
1 ft^3 (air) = 0.08009 lb = 36.5 g at N.T.P.
1 gallon/lb = 0.010 cm^3/kg
1 μg/m^3 = 10-6 g/m^3
1 mmBtu per short ton = 0.325 kWh/kg
(f) Pressure, Pa (Pascal)
1 lb/ft^2 = 4.88 kg/m^2 = 47.88 Pa
1 lb/in^2=702.7 kg/m^2 =51.71 mm Hg=6.894757 10^3 Pa = 6.894757 10^3 N/m^2
1 atm = 1.013 10^5 N/m^2 = 760 mm Hg = 101.325 kPa
1 in H_2O = 2.491 10^2 N/m^2 = 248.8 Pa = 0.036 lb/in^2
1 bar=0.987 atm=1.000 10^{6}dynes/cm^2 =1.020 kgf/cm^2=14.50 lbf/in^2=10^5 N/m^2=100 kPa
1 torr (mm Hg 0°C) = 133 Pa
1 Pascal (Pa) = 1 N/m^2 = 1.89476 kg
1 inch of Hg = 3.377 kPa = 0.489 lb/in^2
1 atmospheric = 1.01325 bar

(continued)

(continued)

(g) Velocity, m/s
1 ft/s = 0.3041 m/s
1 mile/h = 0.447 m/s = 1.4667 ft/s = 0.8690 knots
1 km/h = 0.2778 m/s
1 ft/min = 0.00508 m/s
(h) Force, N
1 N (Newton) = 10^5 dynes = 0.22481 lb wt = 0.224 lb f
1 pdl (poundal) = 0.138255 N (Newton) = 13.83 dynes = 14.10 gf
1 lbf (i.e. wt of 1 lb mass) = 4.448222 N = 444.8222 dynes
1 ton = 9.964 10^3 N
1 bar = 10^5 Pa (Pascal)
1 ft of H_2O = 2.950 10^{-2} atm = 9.807 10^3 N/m^2
1 in H_2O = 249.089 Pa
1 mm H_2O = 9.80665 Pa
1 dyne = 1.020 10^{-6} kg f = 2.2481 10^{-6} lb f = 7.2330 10^{-5} pdl = 10^{-5} N
1 mm of Hg = 133.3 Pa
1 atm= 1 kg f/cm^2 = 98.0665 k Pa
1 Pa (Pascal) = 1 N/m^2
(i) Mass flow rate and discharge, kg/s, m^3/s
1 lb/s = 0.4536 kg/s
1 ft^3/min = 0.4720 1/s = 4.179 10^{-4} m^3/s
1 m^3/s = 3.6 10^6 l/h
1 g/cm^3 = 10^3 kg/m^3
1 lb/h ft^2 = 0.001356 kg/s m^2
1 lb/ft^3 = 16.2 kg/m^2
1 litre/s (l/s) = 10^{-3} m^3/s
(j) Energy, J
1 cal = 4.187 J (Joules)
1 kcal = 3.97 Btu = 12 10^{-4} kWh = 4.187 10^3 J
1 Watt = 1.0 J/s
1 Btu = 0.252 kcal = 2.93 10^{-4} kWh = 1.022 10^3 J
1 hp = 632.34 kcal = 0.736 kWh
1 kWh = 3.6 10^6 J=1 unit
1 J = 2.390 10^{-4} kcal = 2.778 10^{-4} Wh
1 kWh = 860 kcal = 3413 Btu
1 erg = 1.0 10^{-7} J = 1.0 10^{-7} Nm = 1.0 dyne cm
1 J = 1 Ws = 1 Nm

(continued)

(continued)

1 eV = 1.602 10^{-19} J
1 GJ = 10^9 J
1 MJ = 10^6 J
1 TJ (Terajoules) = 10^{12} J
1EJ (Exajoules) = 10^{18} J
1 mmbtu = 1,055,073,600 KJ/Kg, = 295 kWh
1MJ = 0.28 kWh
1 KJ =0.000278 kWh
(k) Power, Watt (J/s)
1 Btu/h = 0.293071 W = 0.252 kcal/h
1 Btu/h = 1.163 W = 3.97 Btu/h
1 W = 1.0 J/s = 1.341 10^{-3} hp = 0.0569 Btu/min = 0.01433 kcal/min
1 hp (F.P.S.) = 550 ft lb f/s = 746 W = 596 kcal/h = 1.015 hp (M.K.S.)
1 hp (M.K.S.) = 75 mm kg f/s = 0.17569 kcal/s = 735.3 W
1 W/ft^2 = 10.76 W/m^2
1 ton (Refrigeration) = 3.5 kW
1kW=1000 W
1 GW = 10^9 W
1 W/m^2=100 lux
(l) Specific Heat, J/kg°C
1 Btu/lb°F = 1.0 kcal/kg°C = 4.187 10^3 J/kg°C
1 Btu/lb = 2.326 kJ/kg
(m) Other conversion (carbon footprint)
1 kg CO_2 per mmBtu = 1/295 kg CO_2/kWh = 0.00339 kg CO_2/kWh
1 gCO_2per mmBtu = kg CO_2/(1000 295)kWh=-0.00000339 kg/kWh
1 MJ/kg = 0.28 kWh/kg
(n) Concentration of gas
1mg/kg = 1ppm (part per billion)
1ppm =1000ppb(part per billion)
1ppb =1ppt (part per trillion)
1ppt =0.000001 ppm
1 g/m3 = 1 mg/L = 1 ppm
1ppm = $\frac{1\mu g}{g}$
(o) Temperature,°C and K used in SI
$T_{(Celcius, °C)} = (5/9)\,[T_{(Fahrenheit, °F)} + 40] - 40$
$T_{(°F)} = (9/5)\,[T_{(°C)} + 40] - 40$
$T_{(Rankine, °R)} = 460 + T_{(°F)}$

(continued)

(continued)

$T_{(Kelvin,K)} = (5/9)\, T_{(°R)}$
$T_{(Kelvin,K)} = 273.15 + T_{(°C)}$
T(°C)= T(°F)/1.8 = (5/9) T(°F)
(p) Rate of heat flow per unit area or heat flux, W/m^2
1 Btu/ft^2 h = 2.713 kcal/m^2 h = 3.1552 W/m^2
1 kcal/m^2 h = 0.3690 Btu/ft^2 h = 1.163 W/m^2 = 27.78 10^{-6} cal/s cm^2
1 cal/cm^2 min = 221.4 Btu/ft^2 h
1 W/ft^2 = 10.76 W/m^2
1 W/m^2 = 0.86 kcal/hm^2 = 0.23901 10^{-4} cal/s cm^2 = 0.137 Btu/h ft^2
1 Btu/h ft = 0.96128 W/m
(q) Heat transfer coefficient, W/m^2°C
1 Btu/ft^{2}h°F = 4.882 kcal/m^{2}h°C = 1.3571 10^{-4} cal/cm^2 s°C
1 Btu/ft^{2}h°F = 5.678 W/m^2°C
1 kcal/m^{2}h°C = 0.2048 Btu/ft^2 h°F = 1.163 W/m^2°C
1 W/m^{2}K = 2.3901 10^{-5} cal/cm^{2}sK = 1.7611 × 10^{-1} Btu/ft^2°F= 0.86 kcal/m^{2}h°C
1 W/m^{2}K = 1 W/m^2°C
(r) Thermal Conductivity, W/m°C
1 Btu/ft h°F = 1.488 kcal/m h°C = 1.73073 W/m°C
1 kcal/m h°C = 0.6720 Btu/ft h°F = 1.1631 W/m°C
1 Btu in/ft^2 h°F = 0.124 kcal/mh°C = 0.144228 W/m°C
1 Btu/in h°F = 17.88 kcal/mh°C
1 cal/cm s°F = 4.187 10^2 W/m°C = 242 Btu/h ft°F
1 W/cm°C = 57.79 Btu/h ft°F
(s) Angle, rad
2π rad (radian) = 360°(degree)
1°(degree) = 0.0174533 rad = 60′ (minutes)
1′ = 0.290888 10^{-3} rad = 60" (seconds)
1" = 4.84814 10^{-6} rad
1° (hour angle) = 4 minute (time)
(t) Illumination
1 lx (lux) = 1.0 lm (lumen)/m^2
1 lm/ft^2 = 1.0 foot candle
1 foot candle = 10.7639 lx
100 lux = 1 W/m^2
(u) Time, h
1 week = 7 days = 168 h = 10,080 minutes = 6,04,800 s
1 mean solar day = 1440 minute = 86,400 s
1 calender year = 365 days = 8760 h = 5.256 10^5 minutes

(continued)

(continued)

1 tropical mean solar year = 365.2422 days
1 sidereal year = 365.2564 days (mean solar)
1 s (second) = 9.192631770 10^9 Hertz (Hz)
1 day = 24 hour = 360° (hour angle)
(v) Concentration, kg/m^3 and g/m^3
1 g/l = 1 kg/m^3
1 lb/ft^3 = 6.236 kg/m^3
(w) Diffusivity, m^2/s
1 ft^2/h = 25.81 10^{-6} m^2/s

Important Constants

Approximate Values of Some Constants in Renewable Energy Sources Which is Multiple of Three

Constants	Actual value	Approximate value
A		
Absorptivity of bare surface		0.3
Absorptivity of blackened surface		>0.9
Altitude of ozone (O_3) layer present in stratosphere	12–25 km	12–24 km
Average heat flux from centre of earth to Earth's surface	0.06 W/m^2	0.06 W/m^2
Average temperature of the earth (≈25 °C)	298 K	300 K
B		
Band gap for silicon	1.16 eV	1.2 eV
Black body temperature of the sun's surface	5777 K	6000 K
Basic convective heat transfer coefficient from outer bare surface with zero wind velocity	2.8	3
Boltzmann constant	1.38×10^{-23} J/K	12×10^{-24} J/K
Broad classification of thermal comfort (physical, physiological, intermediate) parameters	3	3

(continued)

G. N. Tiwari, *Energy, Ecology and Environment*,
https://doi.org/10.1007/978-981-99-3997-8

(continued)

Constants	Actual value	Approximate value
C		
Central core (0–0.23R) temperature of the sun	8–40 × 10^6 K	9–30 × 10^9 K
Convective heat transfer coefficient for air with V as a wind velocity	2.8 + 3V	3 + 3V
Convective and radiative heat transfer coefficient from bare surface to flowing air	(5.7 + 3.8V) W/m^{2}K	(6 + 3V) W/m^{2}K
Cooking time by solar cooker	2–3 h	3 h
Climatic zone inIndia	6	6
D		
Diameter of the sun (2R_S)	1.39 × 10^9 m	1.5 × 10^9 m
Distance of the sun from the earth	1.5 × 10^{11} m	150 × 10^9 m
Diameter of the earth (D = 2R)	1300 0 km	1.5 × 10^6 m
Density of air	1.2 kg/m^3	1.2 kg/m^3
Density of water	997 kg/m^3	990 kg/m^3
Dry biomass in biosphere	250 × 10^9 ton/year	240×10^9 ton/year
E		
Energy generated at centre core of the Sun	90%	90%
Effect of climatic parameters on yield		9–12%
Effective density of states in conduction bands	2.82 × 10^{19} cm^3	27 × 10^{18} cm^3
Emissivity of surface	0.9	0.9
Efficiency of solar cells in standard conditions	15%	15%
Efficiency of PV module with Si-solar cell	12%	12%
Energy contained in visible region	47% (502.6 W/m^2)	48%
Energy contained in infrared region	51.02% (697.4 W/m^2)	51%
Energy contained in ultraviolet (UV) region		
Energy produced in one fusion reaction inside sun	26.7 meV	24 meV

(continued)

(continued)

Constants	Actual value	Approximate value
F		
Fermentation temperature of slurry for biogas production	37 °C	36 °C
Fin efficiency		0.9
Flat plate collector (FPC) efficiency factor (F')	0.7	0.6
Flow rate factor (F_R)	<1.0	<0.9
FPC connected in series		≤ 3
G		
Gas-turbine operates	600–1200 °C	600–1200 °C
Geothermal energy from the earth	300×10^{12} W	300×10^{12} W
H		
Heating concepts (direct, indirect and isolated)	3	3
Heating value of coal	29000 kJ/kg	30000 kJ/kg
Heating value of biogas	20000 kJ/kg	20000 kJ/kg
Heating value of wood/straw	15000 kJ/kg	15000 kJ/kg
Heating value of Gasolene/Kerosene	42000 kJ/kg	42000 kJ/kg
Heating value of Methane	50000 kJ/kg	51000 kJ/kg
High-temperature geothermal well	≥150 °C	≥150 °C
Hydropower system electrical efficiency (Pelton wheel turbine base)	90%	90%
I		
Ideal efficiency of solar still	60%	60%
Intermediate comfort parameters	6	6
Insulation thickness	0.10 m	0.09 m
J		
Junction thickness near n-type semiconductor in Si	0.15 μm	0.15 μm
L		
Latent heat of vaporization	2.3×10^6 J/kg	3×10^6 J/kg
Long wave length radiation from Earth	10 μm	9 μm
Long wave length radiation exchange (ΔR) between ambient and sky	60 W/m^2	60 W/m^2
Low temperature geothermal well	≤150 °C	≤150 °C
M		
Mature tree consumes, CO_2	12 kg of CO_2	12 kg of CO_2
Maximum temperature in concentrating collector		3000 °C

(continued)

(continued)

Constants	Actual value	Approximate value
Maximum wind power extraction	a = 1/3	a = 1/3
Maximum hydropower mechanical factor	$u_t = u_j/3$	$u_t = u_j/3$
Maximum power coefficient (WECS)	59%	60%
Maximum efficiency of WECS with ideal Pelton wheel curved turbine	100%	90%
Mean sun-earth angles (three w.r.t. centre of earth and three w.r.t. observer on earth)	6	6
Methane presence in biogas	60%	60%
N		
Number of sunshine hours (average)	5	6
Number of solar cell in standard PV module for 18V		36
O		
Optimum till angle for maximum solar radiation, degree	$\Phi \pm 15$	$\Phi \pm 15$
Order of radiation heat transfer coefficient	6 W/m^2K	6 W/m^2K
Order of convective heat transfer coefficient between hot plate and water	100 W/m^2K	(90–300) W/m^2K
Overall heat transfer coefficient for glazed FPC single glazed		6 W/m^2K
Optimum temperature for body of human	35–37 °C	36 °C
Optimum depth of water in basin of solar still	0.02–0.03 m	0.03 m
Optimum wind velocity for wind turbine	10 m/s	9 m/s
One photon has energy	3.2883 eV	3 eV
OTEC power generation, ΔT	≥15 °C	≥15 °C
P		
Physical comfort parameters	9	9
Physiological comfort parameters	6	6
Propane in liquefied petroleum gas (LPG)	90%	90%
R		
Radius of Earth (R)	6,500 km	6.5×10^6 m
Rate of evaporation from free water surface ($\dot{q}_{ew}$)	$0.016 \times h_{cw} \times (P_w - rP_a)$	$0.015 \times h_{cw} \times (P_w - rP_a)$

(continued)

(continued)

Constants	Actual value	Approximate value
S		
Saturation current in reverse bias (I_o)	10^{-8} A/m^2	0.1×10^{-9} A/m^2
Sky temperature (°C)	(T_a – 12)	(T_a – 12)
Solar constant	1367 W/m^2	1500 W/m^2
Solar intensity in terrestrial region		900 W/m^2
Solar energy for photosynthesis	30×10^{12} W	30×10^{12} W
Solar energy for wind and wave conversion	300×10^{12} W	300×10^{12} W
Solar energy for hydropower	40×10^{15} W	30×10^{15} W
Solar energy for sensible heating	80×10^{15} W	90×10^{15} W
Specific of water	4190 J/kg °C	4200 J/kg °C
Specific heat of air	1 kJ/kg K	1 kJ/kg K
Short wave length radiation	0.23–2.6 μm	0.3–3.0 μm
Sunshine hour at equator	12 h	12 h
Standard test condition	1000 W/m^2 and 25 °C	900 W/m^2 and 24 °C
Stefan-Boltzman constant	5.67×10^{-8} W/m^2 K^4	60×10^{-9} W/m^2 K^4
Sunshine hour at north pole	24 h	24 h
T		
Thickness of n-type semi-conductor in silicon solar cell	0.2 μm	0.3 μm
Thickness of p-type semi-conductor in silicon solar cell	0.50 mm	0.60 mm
Threshold intensity (a) Winter (b) Summer		>50 W/m^2 >50 W/m^2
The rate of heat generated by healthy person during sleeping	60 W	60 W
The rate of heat generated by healthy person during hard work	600 W/m^2	600 W/m^2
The body temperature of human being	37 °C	36 °C
Thermal conductivity of insulating material (K)	0.03–0.04 W/m K	0.03 W/m K
Transmittivity of window glass	0.9	0.9
The rate of ventilation/infiltration	0.33 NV$(T_r - T_a)$	0.33 NV$(T_r - T_a)$
Tidal power from planetary motion	3×10^{12} W	3×10^{12} W
Thermal solar energy absorbed by earth	120×10^{15} W	120×10^{15} W
Type of radiation (beam, diffuse and reflector)	3	3

(continued)

(continued)

Constants	Actual value	Approximate value
U		
UV solar radiation	0–0.30 μm	0–0.30 μm
V		
Vehicle using gasoline produces	2.5 kg of CO_2/l	2.4 kg of CO_2/l
V group impurity concentration	10^{15} cm^3	10^{15} cm^3
Velocity of light, c	299,792,458 m/s	3×10^8 m/s
W		
Wein's displacement law	$\lambda T = 2897.6$ μm K	3000 μm K
Wavelength radiation from the sun	0–30 μm	0–30 μm

www.ingramcontent.com/pod-product-compliance
Ingram Content Group UK Ltd.
Pitfield, Milton Keynes, MK11 3LW, UK
UKHW020807290926
15470UKWH00003B/1005

* 9 7 8 9 8 1 9 9 3 9 9 6 1 *